Field & Laboratory Exercises
IN ENVIRONMENTAL SCIENCE

SEVENTH EDITION

Eldon D. Enger
Delta College

Bradley F. Smith
Western Washington University

with contributions by:

Heidi Marcum
Baylor University

David A. Aborn
Archbold Biological Station

W. Merle Alexander
Baylor University

Boston   Burr Ridge, IL   Dubuque, IA   Madison, WI   New York   San Francisco   St. Louis
Bangkok   Bogotá   Caracas   Lisbon   London   Madrid
Mexico City   Milan   New Delhi   Seoul   Singapore   Sydney   Taipei   Toronto

McGraw-Hill Higher Education

A Division of The McGraw-Hill Companies

FIELD & LABORATORY EXERCISES IN ENVIRONMENTAL SCIENCE,
SEVENTH EDITION

This book is printed on recycled, acid-free paper containing 10% postconsumer waste.

67890 QPD/QPD 098765 43

ISBN 0–07–290913–7

Vice president and editorial director: *Kevin T. Kane*
Publisher: *Michael D. Lange*
Senior sponsoring editor: *Margaret J. Kemp*
Senior developmental editor: *Kathleen R. Loewenberg*
Marketing manager: *Michelle Watnick*
Senior project manager: *Kay J. Brimeyer*
Production supervisor: *Laura Fuller*
Design manager: *Stuart D. Paterson*
Senior photo research coordinator: *Lori Hancock*
Compositor: *GAC—Indianapolis*
Typeface: *9/11 Times Roman*
Printer: *Quebecor Printing Book Group/Dubuque, IA*

Cover design: *Sean M. Sullivan*
Interior design: *Jamie O'Neal*

Some of the laboratory experiments included in this text may be hazardous if materials are handled improperly or if procedures are conducted incorrectly. Safety precautions are necessary when you are working with chemicals, glass test tubes, hot water baths, sharp instruments, and the like, or for any procedures that generally require caution. Your school may have set regulations regarding safety procedures that your instructor will explain to you. Should you have any problems with materials or procedures, please ask your instructor for help.

www.mhhe.com

Table of Contents

Preface *v*

Part 1   Ecological Principles 1
1   Introduction and Lab Format 3
2   Community Structure 9
3   Estimating Population Size 25
4   Species Diversity 29
5   Habitat and Niche 37
6   Ecological Competition 49
7   Is Your Campus Friendly to Wildlife? 55
    Field Trip Suggestions 59
    Alternative Learning Activities 59

Part 2   Population Growth 61
8   Population Dynamics 63
9   Human Population—Changes in Survival 71
10   Human Population Dynamics 79
    Field Trip Suggestions 87
    Alternative Learning Activities 87

Part 3   Resource Issues 89
11   Water Awareness 91
12   Water Pollution 99
13   Stream Ecology 107
14   Stream Quality Assessment 113
15   Air Pollution 119
16   Soil Management 129
    Field Trip Suggestions 137
    Alternative Learning Activities 137

Part 4   Energy Use 139
17   Economics of Energy Consumption 141
18   Renewable Energy 147
19   The Effectiveness of Insulation 157
20   Personal Energy Consumption 161
    Field Trip Suggestions 168
    Alternative Learning Activities 168

Part 5   Lifestyle Choices 169
21   An Environmental Survey 171
22   Land-Use Planning: A Shopping Center 177
23   Environmental Awareness and Lifestyle 181
24   Our Finite Resources: The Current Affair 185
    Field Trip Suggestions 189
    Alternative Learning Activities 189

Appendix A   Random Numbers Table 191
Appendix B   Transforming e^x to x 193
Appendix C   The Beaufort Scale 195

Preface

The major objectives of a laboratory class, and this manual, are to provide the students with hand-on experiences that are relevant, easy to understand, and presented in an interesting, informative format. This lab manual has been extensively rewritten to provide the student with the latest information and most applicable laboratory activities possible. The manual has been expanded to provide students with more choices in activities that illustrate crucial environmental issues and relevant topics. Further, the expanded choice of labs allows each instructor to select activities that are tailored to the specific needs and circumstances of his or her class. Ranging from field and lab experiments to conducting social and personal assessments of the environmental impact of human activities, the manual presents something for everyone, regardless of the budget or facilities of each class.

These exercises are grouped by categories that can be used in conjunction with any introductory environmental textbook. All lab activities have been field tested over the past six years, and are easy to do within a two-hour time period. Students, regardless of the degree of science background, will benefit from the variety of laboratory activities offered. Relevance sections for each lab have been included to help the student see that he or she is doing more than "just counting trees." The instructions for each lab have been clarified and all sections shortened and updated. Finally, the students themselves have had extensive input on how to best improve the lab activities, ensuring that the voices of those most impacted by the manual are also heard.

Eldon D. Enger
Bradley F. Smith

PART 1

Ecological Principles

To understand the actions and reactions that occur in the environment, we must begin by studying organisms and their interactions with the environment. From this knowledge, we will increase our understanding of nature and human impacts on it. During the past several decades, it has become increasingly clear that any trespasses we commit against nature are also committed against ourselves. With increasing rates of species extinctions and habitat destruction, it is imperative that we learn all we can about the natural world, and the species that inhabit it. As we increase our knowledge, we humans must be willing to make some hard decisions regarding the species and ecosystems that share this planet with us.

Part I introduces some basic concepts of ecological communities, with each lab building upon the knowledge gained from previous ones. Exercise 1 introduces the student to the scientific method and helps develop scientific thought processes. Exercise 2 describes the diversity of community structure and function. Exercise 3 provides techniques for estimating a species' population size, while exercise 4 discusses species diversity. Exercise 5 examines concepts of habitat and niche. Exercise 6 analyzes competition between organisms for shared or limited resources. Finally, exercise 7 addresses whether your own campus provides the necessary resources for survival of wildlife.

EXERCISE 1
Introduction and Lab Format

I. Objectives

After completing this introduction the student will be able to:

1. Understand the nature of environmental studies and laboratory techniques.
2. Apply scientific methods to experimentation in this course.
3. Write a scientific laboratory report.

II. Introduction

The Problem:

Human beings (and all living things) are a product of their environment. We depend completely upon the various systems in nature that provide, clean, and cycle the very air, water, and soil upon which survival balances. Humans have a critical interest in ensuring the sustainable condition of these systems comprising the various environmental compartments. Without air, water, and soil, a person can survive only for minutes, days, or weeks, respectively. Without sufficiently clean air, water, and soil, the quality of life deteriorates to unacceptable levels at some point. Beyond the pragmatics of survival is a moral obligation humans have toward the environment. For the first time in the history of the earth, a single species has acquired the ability, through invention and voluntary actions, to change the global environment. Through these actions, humanity threatens to alter or destroy the balance achieved over millions of years, which created the very conditions allowing not only human evolution, but that of all life as we know it. This acquired ability makes humankind responsible for our actions and requires that we seek a sustainable social presence and a balance between human and ecological needs.

The Solution:

Because we are a part of the problem, we must be a part of the solution. By buying and using products that are not produced with the environment in mind, we are contributing to the problem. By mindlessly allowing harsh chemicals to be flushed down the drain, by burning petroleum products in your car, and by not reducing your household wastes and allowing them to be buried in a landfill, you are contributing to environmental problems. Solutions require first an awareness of the problem, then a determination of the extent of the problem, and finally action to solve the problem. It is the intent of this manual to continually make you, the student, aware of what the problems are in environmental issues and of actions that can be taken to solve these problems. You, through your actions, are the solution, like it or not, to future environmental sustainability. You can help by:

Rethinking your role

Reducing what you use

Reusing what you use

Recycling what you use

Refusing to accept a lesser quality of life for future generations

Environmental Studies

Environmental studies is the interdisciplinary study of the interaction between human beings and their environments. It examines the effects of humanity on other living organisms and on the nonliving physical environment, the sustainability of natural resources, and the environmental impact on human quality of life. Environmental studies is considered to be interdisciplinary because it takes into consideration information not just from classical sciences such as biology and chemistry, but also disciplines such as anthropology, sociology, law, and economics. In this course, you will apply techniques developed in the physical, natural, and social sciences to real-world environmental problems and relate them to humanities' interactions with nature. Many of the methods taught are used throughout the world to analyze and solve environmental problems and will allow you to compare your results to those obtained by professionals in the field.

Scientific Investigation

Science is a methodical and precise way to study the natural world. Science is one tool for solving problems; it simply discovers facts and relationships based on observation. Scientific investigation typically consists of these steps:

1. Make observations.
2. Formulate questions or hypotheses.
3. Design a study or controlled experiment.
4. Collect data.

5. Interpret data.
6. Draw conclusions.

It is often crucial to review the literature in the field to determine if the same or related experimentation has already been done. This prior knowledge, if any, can corroborate or confound data and conclusions from your study. Once step six has been completed, the process can then be repeated until the questions asked about a topic have been answered. Objects or events to be questioned or explained may be found in natural settings or planned as laboratory experiments. The important factor is that for objects or events to be explained, they must be directly observable.

Hypotheses

Once a person has made a series of observations, he or she must do something with the resulting data. Some sort of general statements or hypotheses must be made about the data. One type of generalization summarizes and makes a statement about a set of data. For instance, a person may notice that the sycamore tree in his or her yard begins to lose its leaves in late October. After observing other sycamore trees in neighboring yards losing leaves at the same time, the observer concludes that all sycamore trees begin to lose their leaves in late October. This conclusion is an example of a generalization based on inductive reasoning, or induction. Induction involves reasoning logically from the specific to the general, from isolated observations to a general statement. Generalizations can be drawn confidently only if a large number of observations has been made that reduce the distorting effects of individual differences.

The hypothesis about leaf fall can be tested by observing other sycamore trees. As more information is gathered, the observer may find it necessary to change the original hypothesis. For instance, the hypothesis may have to be changed to: All sycamore trees in the Southwest lose their leaves in late October.

A second type of hypotheses is an explanatory hypothesis. This type of hypothesis goes beyond a simple summary, and attempts to determine the cause(s) or reason(s) for certain observations. One explanatory hypothesis might propose that sycamore trees begin to lose their leaves because of a change in the photoperiod. Another may relate leaf loss to changes in plant hormone levels. These hypotheses are easily tested by setting up experiments comparing leaf loss under various photoperiod conditions, or by altering plant hormone levels. Both hypotheses meet the scientific requirement of being testable. Hypotheses may be stated in "if . . . then" form, for example:

HYPOTHESIS

If sycamore trees begin to lose their leaves in late October because of changes in plant hormone levels, then a large artificially induced change in hormone levels at other times of the year should also result in leaf loss.

Hypothesis testing involves the use of deductive reasoning. A person using deductive logic starts with general observations and makes a specific conclusion. The "if" portion of the statement is the hypothesis, and the "then" portion contains the predictions that are based on the assumptions made in the hypothesis. If after experimentation the predictions prove to be false, then the hypothesis is false. In science, hypotheses can never be absolutely proven. There will always be alternative explanations for why the results turned out the way they did. Through careful research design and by repeatedly disproving hypotheses, we can make stronger conclusions.

CONCLUSION

Science does not deal with certainties, but with probabilities.

Many false hypotheses have been accepted because they led to true predictions. To show that such a hypothesis is false, other tests must be developed. One false experimental result does not always mean the hypothesis must be totally discarded. It is more a question of the percentage of results that support the hypothesis for it to have any value. Statistical analysis is used to determine the significance of any deviations seen during prediction testing. The larger the sample size or the greater the number of observations, the more likely the hypothesis can be accepted or rejected confidently.

Quantitative and Qualitative Data

Data may be quantitative or qualitative. Quantitative data can be expressed numerically and statistically analyzed. Qualitative data are expressed verbally and may incorporate subjective and cultural values. Both qualitative and quantitative data may be collected from experimentation and observation. Scientists prefer quantitative data. This type of data can be presented directly to the reader without subjective interpretation, which may introduce bias. Quantitative data can also be checked easily by other experimenters. Qualitative data are less easily checked and present fewer opportunities for verification than do quantitative relationships.

Data must also be organized in such a way as to give the investigator useful information. If the experiment used control groups, the easiest organization would be to compare data from each group. If similar measurements were made at various locations or under various treatment conditions, the data could be organized by group, location, or type of measurement.

Limitations

Science has some limitations. Scientists cannot study any phenomenon that cannot be observed. Science cannot be used to make moral or value judgments. Science can predict human reactions to a given situation and it can provide information to people so they may make moral and value judgments for themselves.

All students should remember that there are no necessarily "right" answers for the exercises in this lab manual and there is nothing "wrong" with results that contradict predictions or hypotheses. Remember the statement from earlier: Science does not deal with certainties, but with probabilities.

In a chemistry lab, many of the variables that might adversely affect the outcome of an experiment can be controlled (e.g., temperature, humidity, air pressure). Therefore, the student can predict results with confidence and may assume that some human or mechanical error has been made if those results differ from predictions. In this course, however, the outcome of each exercise depends on a multitude of variables that may not as yet be fully understood. Many environmental factors that cannot be controlled by the experimenter are involved. Results may be predicted only in the most general way and may be perfectly correct even if they differ significantly from those predicted. Often, the real value in an experiment is what is discovered from having a hypothesis disproved. Some of the most valuable scientific advancements have come as serendipitous discoveries.

III. Relevance

Students taking a lab course for the first time are often intimidated about science and the process of successfully participating in a laboratory course. Some students may feel that they are not capable of doing "real science," while others simply may not be interested. There is also the lack of understanding about why you may be doing a particular lab. In fact, students in many labs merely "go through the motions" of performing the activity, with no real understanding of the relevance or meaning of what they are doing. This manual strives to make the labs interesting, informative, and relevant. It does a person no good if he or she can measure tree height without understanding the reason for this measurement. This is a particularly important factor when environmental issues are being studied, since humans are directly and indirectly affected by their environment. Citizens must become more informed about the science, and the scientific process, involved in gathering data about pollution, species extinction, population, resource depletion, energy use, and so on. On a larger scale, although environmental studies is based on gathering data by using the scientific method, on top of this are the social, economic, ethical, religious, and political realities of our society. Regardless of what the data tell us, if society is unwilling to accept the information, or to make changes based on data, we will not reach sustainability. So, in order to become better informed people, it is important to understand the process, limitations, and function of science in society.

IV. Activity

Information and Instructions for Lab Class

In General

This course is an introductory-level course that usually counts as a lab-science credit for a variety of majors. Therefore, students in this class come from a wide range of science and nonscience backgrounds. Some of you may be familiar with the concepts presented in class, while others are getting their first introduction to the material. Contribute your knowledge in class discussions, listen to your classmates, and ask questions. You may discover new viewpoints and aspects of what you thought were familiar issues. For those of you who do not have a strong science background, do not panic—you will have just as much opportunity to learn as anyone else. Feel free to participate in class discussions and to ask questions. Never hesitate to ask a question because you think it is silly. Some of the best discussions start from supposedly "silly" questions.

Some students, especially those who are not science majors, may have trouble seeing the relevance of some of the topics that are covered and the exercises that are performed. Each chapter has a section entitled "Relevance," which is designed to show you why each topic is important to environmental issues. You should read this section carefully, and you are encouraged to engage your instructor and classmates in discussions pertaining to the topic you are covering.

Some nonscience majors may dislike science in general, and not see the point of any of the labs. These students should understand that science is an integral part of all our lives. By seeing how science works, nonscientists may hopefully gain a better appreciation of the capabilities and limitations of science.

At the conclusion of an exercise, the instructor will lead the class in a discussion of what has been discovered. Pay close attention to both the written and oral instructions for answering the questions and/or writing reports. Since each unit has slightly different requirements, the instructor will explain how to write your report. Do not leave class without a complete understanding of how to complete the assignments. Be sure to address each of the questions (listed in the Questions section of each chapter) in your lab write-up.

V. Procedure

In Class

Most of the activities and procedures are designed to demonstrate complex concepts in environmental science and, although not difficult to understand, can be quite detailed and involved. The lab instructor is there to guide you through the exercises, to serve as a facilitator, not to familiarize you with the procedures. If you come to class and are not familiar with the procedures to be followed for that day, you will be behind at the outset. You will impede the class if you require the instructor's time to familiarize you with the procedure.

SAFETY CONSIDERATIONS: Before coming to class, you should read and understand the material in the lab manual.

This is the most important instruction you will receive for this lab class.

Being prepared will allow you to focus on the outcome and its relevance to environmental issues. You will get more out of the class this way. There are no shortcuts to environmental solutions.

Each class will begin with a lecture by your instructor, which includes additional information and an explanation of the procedure. It will be helpful if you have read newspapers, magazines, etc. Your participation in all discussions is encouraged and expected. Take careful notes because information presented in the lecture may be necessary to complete the exercise and may be required in your lab write-up.

Take notes during the procedure on anything you observe that could have an impact on the results of the exercise. Record this information on the data sheets. Follow the procedure exactly so that experimental results will be valid. Record your data in a neat and logical format. Come to lab prepared for the activity in which you will be involved. Bring a calculator, lab notebook, and lab manual to each lab. In some labs (water and soil labs), you will handle chemicals, some of which are corrosive. You will be issued safety glasses and gloves, and you should dress in clothing on which you won't mind possibly getting stains and holes. Please leave the laboratories and equipment as you found them when you entered the lab.

In the Field

Field trips are an opportunity to examine up-close the relationships explored in this class. They are designed to better help you understand the interrelationships between humans and ecology by directly observing interactions of plants, trees, animals, conditions, soils, and activities. Depending upon the time of year of your class, the weather may be hot, cold, wet, or windy. Biting insects, poison ivy, and temperature extremes are all likely to be encountered. Dress and prepare accordingly. We suggest that you do not wear expensive or dress clothing, short pants, sandals, open-toe shoes, etc., for participation in outdoor lab sessions.

The Laboratory Report

One of the most important things you will do in this class (and possibly your career) is write laboratory reports. Your report is your tool for expressing what you did, why you did it, and what you learned in the process. Even if your understanding of the procedure, techniques, and results is perfect, and your results error-free, a poorly-written report will not indicate that you really understand what you have done. Writing reports is not difficult if you remember a few guidelines. Normally, scientific reports are divided into the following sections:

1. Abstract

   The abstract should contain a brief summary of purpose, methods, results, and conclusions of the whole experiment. It is generally no longer than three or four sentences. It may be easier after you have written the rest of the report to write the abstract.

2. Introduction

   Write your introduction in such a way that the reader will be interested in reading the rest of your report. The introduction consists of two parts:

   a. An introduction to the topic and its importance to the environment and society. Cite at least one reference [e.g., (Marcum 1991, p. 32)].

   b. A connection between the introduction and why you performed the exercises (this is your hypothesis). End the introduction with specific questions you intend to answer while testing your hypothesis.

   The first item that should be addressed in the introduction is the importance or relevance of what you've undertaken.

   Examples:

   What is ecological competition?

   What is the difference between intraspecific and interspecific competition?

   Why is information about competition useful?

   Why is it important to look at intraspecific and interspecific competition?

   The second item that should be addressed are the hypotheses being investigated in the report. For example, you might state:

   "I (or "We") hypothesize that intraspecific competition is more intense than interspecific competition in species X."

   The part of the introduction preceding your hypothesis should let the reader know why you think your hypothesis is true. After your hypothesis, you may want to state any predictions you have. For example:

   "I/We predict that if intraspecific competition is more intense in species X than interspecific competition, then the distance between individuals of species X should be greater between each other than the distance between individuals of species X and species Y."

3. Materials and Methods

   This section should include a description of the characteristics of the study area (when applicable) and a summary of what was done and what equipment was used. The procedure should be presented in chronological order and in past tense. You should use the scientific names the first time you mention a species (e.g., "The study area was dominated by sycamore trees, *Platanus occidentalis.*"). Thereafter, you only need to use the common name.

   Your methods section should be complete enough so that anyone reading it would be able to reproduce your experiment with nothing but your report to follow.

4. Results

   Tear out your data sheets and include them with this section. The results section consists of two parts:

   a. Original data obtained from using the procedure.

   b. Data derived or calculated from the information obtained during the lab exercise.

The results section contains nothing more than results. It is not a discussion of what you found (this comes next), it is not a conclusion about what you found (this comes after the discussion), but is only the bare-bones reporting of the facts. State the results of the experiment without opinion, interpretation, or explanation. This is the first time, other than in the abstract, that results have been mentioned. All tables and graphs should follow immediately behind this section. Every table and figure should be numbered and referenced in the text. If you include tables or figures that you do not reference, then they are unnecessary. All tables and figures must be labeled with a one-sentence description or title that tells what is being shown. The reader should not have to refer back to the text to understand the information they are looking at. Descriptions of tables go above the table, whereas descriptions of figures are placed at the bottom of the figure.

5. Discussion

This is the most important portion of the report. In this section, you will discuss your data and any trends or relationships that appeared in the data. The first item that should be addressed here is the interpretation of the results. For example, which type of competition is most intense, and which species compete most intensely? Provide insight as to why such trends may have occurred. Secondly, in this section you should discuss why your results support or disprove your hypothesis (was the field technique appropriate; were the counts accurate; what do these results mean for the future of the species in the area; how does this relate to succession, etc.?). Are your results consistent with what you expect? Why or why not? Are your results above or below accepted environmental quality standards? Why or why not? Answer the questions you asked in the introduction section. Any problems encountered during the procedure that may have caused errors should be discussed. Pay close attention to both human error and equipment error.

Whenever necessary, suggest other experiments that should be done, or additional data that should be collected to answer your initial questions (in your introduction) more thoroughly.

6. Conclusion

Conclusions are to be based on data, and they should follow from your discussion. All conclusions drawn should relate to the statement of the problem (your hypothesis). Did your results confirm or deny your hypothesis? Were your results above or below accepted standards? This section can be a brief, three- to four-sentence paragraph, in which you summarize very clearly your conclusions.

7. References, Literature Cited

Use the format employed in your lab manual. Your lab manual should be cited, of course, as well as any other source you have used. In the text of your report, if you cite specific information, or quote data or persons, cite references using the author's surname, year of publication, and page number [e.g., (Enger and Smith 1997, p. 32)]. If referencing other work in general, no page number is required. In the bibliography, use the following citation format:

Enger, E. D. and B. F. Smith. 2000. *Environmental Science: A Study of Interrelationships,* 7th ed. McGraw-Hill, Dubuque, IA. 456 pages.

Helpful Hints:

When writing your report, the following hints may be helpful:

- Do not put off writing your lab report until the night before it is due. Write it as soon as possible after completing the exercise.
- Write in active voice. For example: "I shook the solution until it was thoroughly mixed," rather than "The solution was shaken until it was thoroughly mixed." The pronouns we, he, and she may also be used where applicable; for instance, "As a group, we compiled data from two sources."; "We laid out a 100-meter transect, using a tape measure and string, then . . . "; or "Our results indicate that . . . ". In some cases, it may be necessary to use the passive voice, but it is not a preferred usage.
- Write in past tense unless it is ridiculous to do so, or unless past tense makes the meaning unclear. The use of past tense is a convention of technical writing.
- Put section headings in your report. These make the report easier to grade and will give it a professional appearance.
- Put your data into graphs or tables whenever possible. Be sure to carefully title all graphs and tables. Refer to these by number in the text of your report.
- Strive for a professional appearance in your work. Write as though you are being paid to do so. Neatness, clarity, style, and appearance of lab reports are important.
- Even though you may have worked in groups, write your own lab report.
- Be sure that you have a separate cover sheet, or title page, showing the title, name and number of the lab, your name, and the date.
- Each section of the report (abstract, introduction, etc.) should begin on a new page.
- The report must be typewritten, double-spaced, 12-point Times Roman font, and margins of approximately one inch.
- Charts, tables, figures, and/or graphs should be referred to in the text, not just included, have a legend, and be on a separate sheet of paper. They should be placed as close to their reference as possible.
- Length should be sufficient to fully address all issues and requirements, but no longer than necessary.

VI. Questions

1. What actions do you do that harm the environment? That help preserve the environment?

2. Have you ever had an environmental studies course? What do you expect to get from this course? What are three of the most important issues you would like to learn about in this class?

3. Take any three observations you have made today (it can be about your friends, nature, studying, or anything else), and formulate three hypotheses that attempt to explain your observations.

4. In your own words, summarize the seven parts of a good lab report. Write down any questions you may have about writing a report, so that your instructor may provide written answers to them.

I. Objectives

The student will gather and analyze data on community structure. After completing the work associated with this exercise, the student will be able to:

1. Understand the ecological concept of community structure and the relevance of this concept to a sustainable environmental policy.
2. Use the quadrat method for studying communities.
3. Determine the density and relative density of species in the community.
4. Determine the frequency and relative frequency of each species in the community.
5. Recognize the importance of community structure in the function of an ecosystem.

II. Introduction

Organisms of the same species that live in the same habitat compose a population. A community consists of an assemblage of populations existing in a common area, which interact with each other and share the same general resources. Plants, animals, and microorganisms are all part of a community. Abiotic components of the environment, such as precipitation, temperature, and soil help define that community. Some communities, such as the arctic tundra, are relatively simple, with only a few hundred species interacting. Others, like tropical rainforests, contain many thousands of species, which produce a complex web of interactions.

Humans can learn a great deal by studying communities. Because we are a part of nature, we must learn to live within the limits imposed upon us by our natural communities. Many of our activities, such as urban development and modern farming practices, have had unforeseen effects on the structure and function of communities. By studying both natural and disturbed communities, we can learn to minimize undesirable effects.

Communities are incredibly diverse, both in structure and function. It would be impossible to study every organism in a community, but we can study samples, and gain insight into the function of the entire community. In a community, the data that researchers gather may not encompass every species and individual in that community, but researchers can use this data to approximate the entire community.

To get a more complete picture of a community, other procedures, such as aerial photography and analysis of important abiotic factors (i.e., erosion) are also included in ecologists' studies. Studying communities involves gathering a great deal of data; however, this exercise will cover only the first procedure usually performed by ecologists, which is surveying community structure. This lab focuses on surveying a grassland-type community, but other communities are easily surveyed by slightly changing the way that species are counted. The lab instructor will be able to help you determine which type of sampling method is most appropriate for your area.

III. Relevance: Keystone Species

As mentioned in the introduction, ecological communities are a complex web of interactions among species. Not all interactions, however, are equal. That is, some species have more of an influence on community structure than others. If one species, say an ant, is removed from a community, the impact may be minimal because there are many other insect species that can fill the "role" of that species. On the other hand, removal of another species may be felt throughout the entire community. These species are referred to as **keystone species** because their presence can determine what species are present in a community and in what numbers. We often think of top predators, such as tigers or sharks, as keystone species, but that need not be the case. Consider krill, the small shrimplike crustaceans that are the primary food of many baleen whales, penguins and other seabirds, and many fish. Without krill, these species might very well go extinct because they rely so heavily on krill. Elephants are another nonpredatory keystone species. By pushing down trees to feed on the upper branches and bark, elephants keep the savannah from becoming a forest. That one species is largely responsible for maintaining an entire ecosystem. Without elephants, the savannahs would eventually become wooded, eliminating many of the herbivores, such as impala and wildebeests. As these species disappear, lions, cheetahs, and hyenas would also disappear, and forest animals would increase.

IV. Activity

This activity quantifies the vegetation structure of a grassland community; therefore, the quadrat method is the most appropriate to use. The quadrat method uses plots of fixed size (usually one square meter) and shape as sampling units. Plots can be square, round, or rectangular. In gathering data, many small samples are more informative than a few large ones. The total sampling size should be at least 10 percent of the study area to accurately predict community structure. For example, if you decide the study area will be 100 square meters, you need to sample at least ten 1-square-meter plots.

The position of plots surveyed should be determined randomly to eliminate bias. Several methods are used to randomly select plots, such as consulting random number tables, as is found in Appendix A of this manual, throwing a stick over one's shoulder, or drawing numbered chips out of a bag. Ecologists record field observations using various methods. Some methods require removing the plants from the field, but we will record the plant species and location on a vegetation map of the community.

A primary function of surveying community structure is determining the abundance of species in relation to habitat, time, each other, or different community types. The measures of abundance are density, frequency, and dominance. **Density** is defined as the number of individuals per area sampled. **Relative density** is the density of a given species in relation to the total density of all species. **Frequency** is the number of quadrats in which a species occurs divided by the number of quadrats examined. **Relative frequency** is the frequency of a given species in relation to the total frequency of all species. From relative density we can determine which species is most abundant. Remember, density measures how many of a species are present, while frequency measures how often that species occurs.

Determining the frequency can help researchers define the distribution of species in a community. Relative frequency can tell us if a given species is distributed randomly, uniformly, or in clumps. If the relative frequency of a species is between 0–30 percent, the species occurs in clumps. If it is between 31–80 percent, the species is randomly distributed, and if it is between 81–100 percent, the species is uniformly distributed throughout the community. It is just as important for managers to know how species are distributed throughout a community as how many individuals of each species are present.

The chart below summarizes the formulas you will use in this lab:

$$\text{Density} = \frac{\text{Number of individuals}}{\text{Area sampled}}$$

$$\text{Frequency} = \frac{\text{Number of quadrats in which a species occurs}}{\text{Number of quadrats examined}}$$

$$\text{Relative density} = \frac{\text{Density of a given species}}{\text{Density of all species}} \times 100$$

$$\text{Relative frequency} = \frac{\text{Frequency of a given species}}{\text{Frequency of all species}} \times 100$$

V. Procedure

Materials

50 m tape or meterstick

1 m² quadrats

marking flags

notebook

data sheets

field guide of local grassland vegetation

random number chart

Method

(This procedure describes sampling in a grassland environment. If a grassland is not available in your area, other community types may be used with slight modification of procedure.)

In Class:

1. Divide the class into groups of 2 to 3 students. Each group will count the plants in 2 quadrats in the field and will provide the rest of the class with the information when everybody is back from the field.
2. The instructor will show the class five major plant species that are to be found in the community you will study. The class will determine which species will be Species a, Species b, and so on, and will record them on Data Sheets 2.1 and 2.2. (Be certain the entire class uses the same letter to represent the same species.)
3. Each group will determine the location of its 2 quadrats by using the random number table found in the back of your lab manual. One member of each group should close their eyes and point to the random number table. The two numbers pointed to will represent the number of steps taken for the first quadrat sample. These first two numbers should be recorded on Data Sheet 2.1. The first number indicates the number of steps forward they should take; the second number indicates the number of steps to the right they should take. Repeat this procedure and record the numbers under the appropriate heading on Data Sheet 2.2.

In the Field:

4. Define the total study area by measuring the desired area (i.e., for our 100-square-meter plot, measure 10 meters to a side and mark each corner with a flag; using this size as an example, a minimum of 10 quadrats will have to be surveyed).

5. Starting at any edge along the study plot, each group should look at the first pair of random numbers it recorded on Data Sheet 2.1. Once these numbers have been walked off, they will lay the quadrat down at their feet. This is the first area they will sample.
6. Count all individual plants belonging to the five species listed at the top of Data Sheet 2.1. Record species type, size, and location on the vegetational community map on Data Sheet 2.1. Indicate a plant's size by drawing a circle proportional to the size of the plant in the quadrat. Record the species using the corresponding letter in the data sheet. What you will end up with is a "map" of the quadrat, showing the relative size and numbers of the five species you are counting. While counting, you must make several decisions. To include a plant in the count, at least 50 percent of it must be in the quadrat. For bushes and grasses, you must make a decision as to what is an individual plant, since one plant may be bushy and may look like several plants, when in fact it is not (your instructor will help you on this).
7. Repeat steps 5 and 6 for your second quadrat, and record results on Data Sheet 2.2.

Back in the Classroom:

8. Count the number of plants of each species in your 2 quadrats.
9. Put your results on the chalkboard and then record the rest of the class results on Data Sheet 2.3, section E. Complete the calculations on Data Sheets 2.4 and 2.5. Answer questions in Section VI.
10. Your instructor will let you know if a report is required for this laboratory.

Data Sheet 2.1: First Quadrat

Name _____

Section _____

Date _____

Location of community:

A. List five species in your community: _____

    Species a: _____

    Species b: _____

    Species c: _____

    Species d: _____

    Species e: _____

    No. of forward paces: _____   No. of right paces: _____

B. Vegetational community map: draw the species present, according to step number 6 under "Procedure."

Data Sheet 2.2: Second Quadrat

Name _____

Section _____

Date _____

Location of community:

C. List five species in your community: _____

    Species a: _____

    Species b: _____

    Species c: _____

    Species d: _____

    Species e: _____

    No. of forward paces: _____   No. of right paces: _____

D. Vegetational community map: draw the species present, according to step number 6 under "Procedure."

Data Sheet 2.3

Name _____

Section _____

Date _____

E. Each group in the class will record the number of plants counted by species in each quadrat surveyed in the community.

Quadrat #	Species a	Species b	Species c	Species d	Species e	Total
1						
2						
3						
4						
5						
6						
7						
8						
9						
10						
Total						

F. Calculations:

1. To find the total density (remember the definition of total density) for all plants in all quadrats:

   a. Add up the total number of plants: _____ plants

   b. Add up the total number of quadrats: _____ sq. meters

   c. Divide a by b to find the total density = (a/b) = _____ plants/sq. meter (put this number in all of the rows in column 4 in the table below)

2. To determine the relative density (remember the definition) of the 5 species (i.e., the density of a species relative to the other species):

   a. Fill in column (1) for the five species according to Data Sheet 2.3.
   b. Column (2) (the no. of quadrats) will be all the same number, which is probably 10.
   c. For column 3, divide the total number of plants (column 1) by the total number of quadrats (column 2) for each species.
   d. Put the answer from (1c) in column 4; again, each row in this column will have the same number.
   e. For column (5), divide column (3) by the total density (column 4) to get a decimal density number.
   f. Multiply each row by 100 (column 6) to convert the decimal density to a percent relative density (hint: the sums of the relative densities should be very close to 100%).

Species	(1) Total no. of plants	(2) Total no. of quadrats (1) ÷ (2)	(3) Species density/m^2	(4) Total density	(5) (3) ÷ (4) /100 m^2 (5) × 100	(6) Relative density
a						
b						
c						
d						
e						

Data Sheet 2.5

Name _____

Section _____

Date _____

3. To determine the total frequency for all the quadrats sampled:
   a. How many quadrats had plants: _____
   b. How many quadrats were surveyed: _____
   c. Divide a by b to find the total frequency = (a/b) = _____ (put this number in all of the rows in column 4 of the table below).
4. To determine the relative frequency of each plant species:
   a. In column (1), for each row, record the number of quadrats that had that particular species.
   b. In column (2), record the total number of quadrats surveyed (again, all rows will probably be 10).
   c. For column (3), divide column (1) by (2) to get a species frequency.
   d. Column (4) will have the answer calculated in 3c.
   e. For column (5), in each row, divide the species frequency (column 3) by the total frequency (column 4).
   f. For column (6), multiply each row by 100 to change the decimal frequency of each species to a relative frequency.

Species	(1) Number of plants species occurs in	(2) Total no. of quadrats	(3) Frequency of species (1) ÷ (2)	(4) Total frequency	(5) (3) ÷ (4)	(6) Relative freq. (5) × 100
a						
b						
c						
d						
e						

VI. Questions

1. According to the vegetational community map, which species appeared to be most dominant? Did all quadrats contain at least one species of plant? Were any quadrats devoid of plant life?

2. Based on the relative density of each species, which plant is the most dominant? Does this correspond with your answer in question 1?

3. Based on the relative frequency of each species, which plant occurs most frequently? Is this the same species that was most dominant?

4. According to the relative frequency of each species, are the plants in the community uniformly distributed, randomly distributed, or clumped in groups? (Answer for each species and as a whole community.)

5. Are there any features or characteristics of the environment (e.g., slope, soil moisture, shade, etc.) that might lead to the distributions you observed?

6. What difficulties could you encounter in trying to run quadrat studies on animal populations? In an aquatic environment?

7. What is the relation between density and frequency? Can a species of plant have a high relative density, but have a lower relative frequency, or vice versa?

8. What significance does calculating density and frequency have in conserving and protecting ecosystems?

9. Have humans influenced the community you studied? If so, how?

EXERCISE 3
Estimating Population Size

I. Objectives

This lab examines ways in which scientists estimate population size. By the end of the exercise students should:

1. Understand the purpose of estimating the size of a population.
2. Understand the mark-recapture method.
3. Be able to estimate the size of a population in their area.
4. Become aware of the reasons for species declines.

II. Introduction

You may have heard or read about species declines, especially with regard to endangered species. Indeed, precipitous population declines are one of the main reasons why species are listed as threatened or endangered. In order to know whether a species population has declined, however, scientists must know the size of the population to begin with. Chances are that unless the population is small to begin with, or all the individuals are restricted to a small area (e.g., an island), it will be impossible to count every individual in a population. Therefore, scientists often have to **estimate** population size, based on a sample. In the community structure lab, you used quadrats to estimate the density and abundance of several plant species. Plants, however, do not move and are therefore relatively easy to sample. So how do scientists sample more mobile species? One common way is to use a technique called **mark-recapture**. It is based on the idea that if you mark a certain number of individuals, release them back into the population, then later resample the population, the proportion of marked individuals you find is equal to the proportion of the total population you marked originally. This may sound confusing, but consider the following example:

Suppose a population of beetles in an area has 1000 individuals. A scientist catches 100 beetles (10% of the whole population), marks them with a small drop of red paint, and lets them go. The next day she returns and captures another 100 beetles, and out of this second sample she finds 10 beetles that were marked the day before (10% of the sample taken). So,

$$\text{Total population} = \text{No. of individuals marked initially} \times \frac{\text{No. of individuals in the 2nd sample}}{\text{No. of marked individuals recaptured}}$$

Using the numbers above,

$$\text{Total population} = 100 \times \frac{100}{10} = 1000 \text{ (a pretty good estimate!)}$$

There are several assumptions that must be met in order for this formula to work:

1. All samples must be random (you can't go looking preferentially for marked individuals).
2. Marked individuals must be given sufficient time to mix back into the population.
3. Marked and unmarked individuals must have an equal chance of capture.
4. Marks must not become lost or unrecognizable.

The formula is not perfect, but it can give a good estimate to work with, as long as the conditions listed above are met.

III. Relevance: The Disappearance of the Songbirds

There was a time when almost any wooded area in the United States, even urban parks, would be filled with vibrant colors and exquisite songs of warblers, vireos, thrushes, orioles, and flycatchers. But starting in the 1950s and 1960s, birdwatchers in some areas noticed that there did not seem to be as many songbirds as there used to be. When such reports started becoming more widespread, the U.S. Fish and Wildlife Service investigated. Much of this work has involved sampling and monitoring populations of species nationwide to estimate population changes over time. In some cases the results are startling. Since the 1970s, Baltimore orioles have declined 30 percent, wood thrushes have declined 40 percent, and golden-winged warblers have declined 46 percent! Some species are showing declines in some parts of the country, but are doing fine in other areas, and some species' populations are steady everywhere. Most, however, are showing some degree of decline. The causes of the declines are many—habitat loss on the breeding and wintering grounds and along migratory routes, increased nest predation from urban and domestic animals, and pesticides, to name a few. Efforts are underway to try to stop and ultimately reverse the declines before many songbirds reach endangered status. Why should we be concerned about songbird populations? Most songbird species migrate to Mexico, Central and South America, and the Caribbean. How can we persuade foreign countries to undertake conservation efforts? Do we have a right to try?

IV. Activity

You will be using the mark-recapture method to estimate the population of insects in your area. Your instructor will take you to an area, such as a local park, where pitfall traps have been set. A pitfall trap is simply a small jar or cup set into the ground. Insects fall into the trap and cannot get out. Your instructor will pick a particular species of insect, such as a beetle or cricket, to look at. A certain number will be sampled and marked. You will then resample the traps and record the number of marked individuals you find to estimate the total population size. Once you are finished with the lab, remember to release all insects back into their habitat.

V. Procedure

Materials

    cups or jars for pitfall traps

    data sheets

    calculators

Method

In the Field:

1. Your instructor will inform you as to which species you will be examining.
2. At least one day prior to lab, a certain number of individuals will have been marked with a small drop of nontoxic paint, then released. Your instructor will inform you how many individuals were marked.
3. Divide into groups and check the pitfall traps for the species you are looking at. Record the total number of individuals your class found on Data Sheet 3.1. Then record how many of those were marked.

Note: If no suitable sampling areas exist in your region, or if the lab is conducted during winter months, you can use mealworms or flour beetles in a container such as an aquarium or large bucket. Each student can sample by sweeping a small sieve through the container and count marked and unmarked individuals. This alternative offers the advantage of the instructor knowing exactly how many individuals are in the total population.

In the Classroom:

1. Calculate the population estimate using the formula presented in the Introduction.
2. Answer the questions in Section VI.

Data Sheet 3.1

Name _____

Section _____

Date _____

Number of individuals originally marked _____

Total number of individuals captured in the resample _____

Number of individuals in the resample that were marked _____

Population estimate _____

VI. Questions

1. How accurate do you think your population estimate is? What are some factors that might have reduced its accuracy?

2. How might you use the estimate you calculated?

3. Do you think this technique would work for other animals, such as fish or birds? Explain.

4. Based on the area where you sampled, would you expect to see population increases or declines if you sampled in future years (i.e., does there appear to be a lot of habitat loss, development, landscaping, etc.)?

Species Diversity

I. Objectives

The student will gather and analyze data pertaining to species diversity in areas of differing human disturbance. After completing this exercise the student should:

1. Understand the difference between species richness and evenness.
2. Be able to calculate species richness and evenness.
3. Understand how natural and human disturbance can affect species diversity.

II. Introduction

Natural ecosystems contain a wide variety of species. These species interact to form stable and functional communities, as you have seen previously. The numbers and proportions of different species that can be found in an area is what we term **ecological diversity**. There are two components that are used to measure the diversity of an area. **Species richness** is simply the number of different species that are found. **Species evenness** is the relative proportion of each species. For example, if 100 individuals from a community are sampled and we find 20 different species, we might say that the community has a high species richness. But if out of those 100 individuals, 81 of them are of one species and there is one individual each of the remaining 19 species, then we would conclude that evenness is low because the community is dominated by just one species. When communities are disturbed, especially as a result of human activity, species richness and/or evenness can decrease. These alterations can change the entire structure and functioning of communities. Therefore, measuring species diversity and changes in diversity is a good way to measure the health of an ecosystem.

III. Relevance: Biodiversity

One of the major concerns of ecologists and conservation biologists is the maintenance of biodiversity. Biodiversity may be defined as the variability that exists within species, communities, and ecosystems. At the community level, the biotic and abiotic interactions function in a way that maintains continuity of the local environment. Properly functioning communities provide humans with many benefits, including flood and erosion control, air and water purification, and consumable resources. As humans disrupt community interactions, usually through habitat alteration and the addition/removal of species, the ecological processes degrade, impacting not only the environment, but humans as well. For example, by destroying grasslands and forests for agriculture, humans have removed the ability of those communities to hold water, as well as increasing soil erosion. As a result, $9,000,000$ km^2 of the Earth's surface that was once grassland, savannah, and forest has become desert, unfit for either native or non-native organisms. With the loss of biodiversity in these areas, humankind has lost economic benefits, potential cures for diseases, and the simple aesthetic beauty of these communities and the species that inhabit them.

IV. Activity

In this activity you will measure plankton species diversity in disturbed and relatively undisturbed aquatic habitats. Since plankton are at the base of many food chains, plankton diversity is one way of examining the health of an aquatic system. To measure diversity, you will use the **Shannon-Wiener Index**. This is a commonly used index that takes into account both species richness and evenness. The formula for this index is:

$H = \Sigma\ p_i \ln p_i$, where p_i is the proportion of species I in your sample, and $\ln p_i$ is the natural logarithm of species I.

Using the example given in the Introduction, the proportion of Species 1 is 0.81 (81 individuals of Species 1 found out of 100 sampled), and the proportion of each of the remaining species is 0.01. To get lnp for each species, enter the number for p, then press the "lnx" button on your calculator. Thus, $H = (0.81)(-0.21) + (0.01)(-4.61) + (0.01)(-4.61) \ldots (0.01)(-4.61) = 1.05$ (it's OK to drop the negative sign). By itself this number doesn't mean much, but when compared with the index for other areas it can tell you whether diversity is higher or lower. The higher the value of H, the higher the diversity, either because of greater species richness, evenness, or both.

V. Procedure

Materials

jars or bottles for collecting water samples

eye droppers

depression slides

compound microscopes

calculators with "lnx" capabilities

Method

In the Field:

1. Your instructor will take you to two aquatic systems (streams, ponds, creeks, etc.), one that is relatively pristine (e.g., in a park, away from the city, etc.) and one that is influenced by human activity (e.g., near a construction site, near a roadside, etc.).
2. At each location, describe the area using the questions on Data Sheet 4.1.
3. At each location you will collect water samples in the containers provided. (*Note:* plankton nets, if available, may be substituted for bottles.) Seal the containers and label them as to which area the sample came from (disturbed or undisturbed).

In the Lab:

1. Divide into pairs, with one member of each pair examining a sample from one area.
2. Using an eye dropper, each student should take a few drops of the water sample and place it on a depression slide. A cover slide is not necessary.
3. Place the slide on a compound microscope and record the different species you see and how many of each you find on Data Sheets 4.2 and 4.3. You do not need to specifically identify each organism, just each different type. For example, if the first plankton you see is green and round, that is Species 1, and each time you see a green, round plankton, you record that as another individual of Species 1.
4. Once you have scanned the entire slide and have recorded the number and relative abundance of each species, rinse off the slide and set it to dry, then calculate the Shannon-Wiener Index.
5. Exchange data with the other member of your pair and answer the questions in Section VI. Your instructor will inform you if a report is required for this laboratory.

Data Sheet 4.1

Name _____

Section _____

Date _____

Answer the following questions about the two areas sampled.

1. Describe your impression of the biotic environment (density, ground cover, maturity, etc.).

   Area 1: _____

   _____

   _____

   Area 2: _____

   _____

   _____

2. Describe the topography of the habitat (slope, location, etc.).

   Area 1: _____

   _____

   _____

   Area 2: _____

   _____

   _____

3. Record other details differentiating the habitats.

   Area 1: _____

   _____

   _____

   Area 2: _____

   _____

   _____

Data Sheet 4.2

Name _____

Section _____

Date _____

Area: _____

Type of organism (short description or sketch)	Number of individuals	Relative abundance (p_i)	$\ln p_i$	$p_i \ln p_i$
	Total =			H =

Data Sheet 4.3

Name _____

Section _____

Date _____

Area: _____

Type of organism (short description or sketch)	Number of individuals	Relative abundance (p_i)	ln p_i	p_i ln p_i
	Total =			**H =**

VI. Questions

1. Was there a difference in plankton diversity between the two areas? If so, what might be responsible for the difference?

2. Were there some types of plankton (e.g., phytoplankton) that seemed to be more abundant in one area than in the other? Why might this be so?

3. How might any differences in plankton diversity you observed affect the aquatic communities you sampled?

EXERCISE 5

Habitat and Niche

I. Objectives

The student will gather and analyze data from forest habitats concerning niche characteristics. After completing the work associated with this lab, the student will be able to:

1. Understand habitat and niche concepts as they relate to community structure and to ecosystems.
2. Determine niche relationships of tree species in the field.
3. Observe niche breadth and niche overlap between tree species in different habitats.
4. Recognize some environmental factors that can affect niche breadth and overlap.
5. Recognize that humans play an important role in constricting or expanding other species' niche breadth and overlap.

II. Introduction

One important factor affecting the distribution of organisms is their habitat. **Habitat** is the region where a plant or animal naturally lives and can normally be found. Many plants and animals occupy the same habitat. Because of this, they must share common resources. **Niche** is the role an organism plays in its community. A niche is part of the set of relationships between a species and its environment, and is much more than the habitat in which that species lives. A shark lives in oceans, that is its habitat. Part of the niche it occupies is that of predator. Other variables in a niche include territory, feeding habits, breeding habits, competition, and physiological constraints. Many species can coexist in a community because they occupy different niches. The **fundamental niche** of a species is the set of *all* environmental conditions that permit it to exist. A person cannot directly observe a fundamental niche because it is more of a concept, defined by an infinite number of variables. Even if a species has the potential to exist in a particular habitat, however, it may not. That portion of the fundamental niche that is *actually occupied* is called the **realized niche**. Competition and other species interactions often determine what portion of the fundamental niche becomes the realized niche. A species' realized niche consists of that species' behavior, physiology, as well as its physical habitat. The spatial or physical components of a niche (i.e., soil moisture, elevation) define the species' habitat.

Niche breadth is defined as the diversity of resources used by a species. It can be used to measure the distribution of species among habitats. Some organisms are able to live in almost any habitat (i.e., rats, cockroaches), while others are severely restricted in their resource use (i.e., pandas and koalas eat only one type of vegetation, a narrow niche). **Niche overlap** is the extent to which species share the same resources. The **competitive exclusion principle** states that no two species can share exactly the same niche for very long, so species in a community that overlap to some extent must use at least some resources differently. For example, a great horned owl and a red-tailed hawk may share the same food source, (rodents, rabbits, etc.) but hunt at different times of the day. Two plants may share the same soil type, but have different methods of pollination (i.e., wind, insects). If two species do overlap, natural selection will eventually result in either the coexistence of both species or the extinction/relocation of one of them.

Species diversity also affects niche breadth and overlap. A community that supports a large number of species correspondingly has a large number of niches. Stability is simply the ability to resist change. Change is less likely to affect every niche in a diverse community than it is likely to affect a community with only one or two species. A single plant disease may have little effect in a tropical rainforest, but it may completely destroy a cultivated monocrop field.

III. Relevance: Habitat Loss

Probably the greatest threat to biodiversity comes from habitat loss. Some countries have lost more than 75 percent of their wildlife habitats. While all habitats are being affected, forests and wetlands are seeing some of the fastest losses. Humans need resources, but as human populations increase, more space is needed and more resources are used up. The land that habitats occupy is needed for development, but habitats also provide benefits (e.g., forests are important to any industry using wood; wetlands are important for fisheries, and also purify our water). What should happen when economic interests conflict with ecosystem interests? Are we morally bound to protect ecosystems that are currently headed for destruction? Is it right to invoke legislation, such as the Endangered Species Act, to stop habitat loss?

IV. Activity

In order to help you understand the importance of habitat and niche and how these fit into our study of ecology and into the larger picture overall of human interaction with the environment, we will examine several tree species in local communities. The niche concept is abstract and impossible to measure exactly. Ecologists have devised methods of calculating values of various indices for niche breadth and niche overlap. Niche breadth and overlap are commonly measured by diversity indices (i.e., the Shannon Index) and statistical methods. However, we will limit our study to observing how several species of plants are proportionally distributed in different habitats.

37

In this activity we will survey tree species in three different habitats. We will record only the type and number of trees; we will not map their location. We can observe niche breadth and overlap of the species in the different habitats by calculating the mean of the three habitats and then entering all the data on a graph. We will be able to tell which species have broad niches (i.e., they will exist in all 3 habitats), which species have narrow niches (they may occur in only 1 or 2 habitats), and which species overlap (they both occur together in two or three of the habitats).

V. Procedure

Materials

4 pieces of surveyor's ribbon (about 18" each)

4 colored pencils

50 m tape measure

data sheets

calculators

Method

In the Classroom:

1. The instructor will describe three tree species that are common in your area and that occur in at least one of the three habitats. Each species counted will be identified by a letter. Record the letter and corresponding species on Data Sheet 5.2.
2. Divide the class into three groups. Each group will be assigned one of the species, and will be responsible for counting only the trees of that species at each habitat.
3. The instructor will give directions to three wooded communities containing diverse habitats (i.e., different types of hillside, river bottom, plateau, soil types, etc.).

In the Field:

4. At the first community, each student will describe the environment by answering the appropriate questions on Data Sheet 5.1. The instructor will help you to identify the tree species in the community, if they haven't already been identified in the lab.
5. Measure a 30-meter by 30-meter square plot and mark the corners with the surveyor's ribbon.
6. Each group is responsible for identifying every tree of its assigned species within the 30 × 30 plot and recording this number on Data Sheet 5.2, section C. Be sure that all members of your species in the plot are recorded, but do not count seedlings or saplings less than 2 inches in diameter.
7. Repeat steps 4 through 6 for the next two habitats and record all data.

Back in the Classroom:

8. Each group will put its results on the board. The class will record these results on Data Sheet 5.2, section D and calculate the average for each species.
9. Graph the number of each species on Data Sheet 5.3 using the following directions:
   a. Use different colors or shading to plot Species a, b, c.
   b. We will use an example to understand how to plot the graph. Say Species a was denoted with black bars, and there were 12 individuals in Habitat 1, 22 individuals in Habitat 2, and 5 individuals in Habitat 3. Species b, denoted by diagonal bars, had 10 individuals in Habitat 1, 5 in Habitat 2, and 30 in Habitat 3, and Species c, denoted by gray bars had 8 individuals in Habitat 1, 12 in Habitat 2, and 3 in Habitat 3. Graphically, the distribution would look like figure 5.1.
   c. From looking at figure 5.1, we can observe niche breadth and overlap in the following ways:
      TO FIND NICHE BREADTH: LOOK AT ALL THREE SITES: (1) species that deviate most from the average have narrow niches. Although this may sound backwards, the answer is quite simple: when a species exists in only one habitat, the average will be skewed (i.e., number + 0 + 0 divided by 3), and the number of that species may be quite far away from the calculated average. The above example shows Species b had a narrower niche than the other two species. If you look at the three sites, you can see that the majority of individuals of Species b are found in only one habitat. It might be termed a habitat "specialist"; (2) species closest to the average have broad niches: if a species exists in all three habitats in about the same numbers, the average for that species will be close to the number found in each habitat. In the above example, Species c occurs evenly in all three habitats; it is what would be called a "generalist."
      TO FIND NICHE OVERLAP: LOOK AT EACH SITE SEPARATELY TO DETERMINE THE AMOUNT OF OVERLAP IN EACH SITE: (3) different species with approximately the same numbers at that habitat have some niche overlap. In the above example, all species share all habitats, with the greatest overlap in Habitat 1.

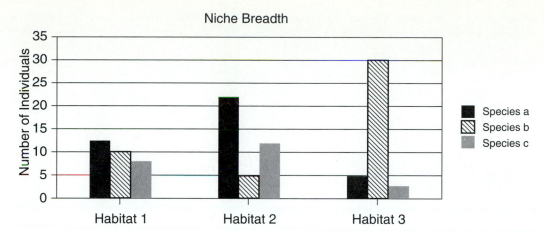

Figure 5.1   Niche breadth

Data Sheet 5.1:

Name _____

Section _____

Date _____

Habitat (to be completed in the field)

Group Number: _____

A. Answer the following questions about the three habitats surveyed.

1. Describe your impression of the biotic environment (density, ground cover, maturity, etc.).

   Habitat 1 _____

   _____

   Habitat 2 _____

   _____

   Habitat 3 _____

   _____

2. Describe the topography of the habitat (slope, location, etc.).

   Habitat 1 _____

   _____

   Habitat 2 _____

   _____

   Habitat 3 _____

   _____

3. Describe the soil (color, texture, organic content, wetness).

   Habitat 1 _____

   _____

   Habitat 2 _____

   _____

   Habitat 3 _____

   _____

4. Record other details differentiating the habitats.

   Habitat 1 _____

   _____

   Habitat 2 _____

   _____

   Habitat 3 _____

   _____

Data Sheet 5.2

Name _____

Section _____

Date _____

Group Number: _____

B. What are the three species the class will count in the field? (Determine in class prior to going into the field.)

      Species a _____

      Species b _____

      Species c _____

C. Count the number of trees in your group's assigned species (only one tree species per group).

    1.   Habitat 1—number of your species _____

    2.   Habitat 2—number of your species _____

    3.   Habitat 3—number of your species _____

D. Tree Species Data Chart—record the total number of trees of each species in each habitat that the class counted (complete in class).

Species	Species' Name	Habitat 1	Habitat 2	Habitat 3	Total (T) (1 + 2 +3)	Average (T/3)
a						
b						
c						

Data Sheet 5.3

Name _____

Section _____

Date _____

Group Number: _____

E. Using data from the table in D. on Data Sheet 5.2, graph the number of trees in each species for each habitat and the species averages. Remember the instructions from the Procedure section 9a.–9c. and use a different color or pattern for each bar on the graph.

F. Using the data from the above graph, answer the questions that follow. Your instructor will inform you if a report is required for this lab.

VI. Questions

1. How do the environmental factors in the different habitats described on Data Sheet 5.1 affect the number and types of tree species in each habitat? What about the density of trees in each habitat?

2. Based on Data Sheets 5.2 and 5.3, which tree species have the narrowest niches? Broadest niches? How did you come to these conclusions?

3. Which tree species show some niche overlap? Explain how you came to this conclusion.

4. Describe some environmental management techniques that you might use to preserve a species with a narrow niche. What about a species with a broad niche?

5. Which type of species (generalists or specialists) do you think would be the hardest to protect? Give an example of a species that humans are currently trying to save. What are some problems with its niche requirements?

6. How does the human niche affect niche breadth and overlap of other species?

7. Can you give any examples of current problems where human intrusion into other species' niches has driven that species to an endangered status or to extinction?

Ecological Competition

I. Objectives

This chapter addresses the ecological implications of competition within a community. For this lab, the student will:

1. Gain an understanding of the concept of ecological competition.
2. Use several methods to observe competition in a bird community.
3. Examine basic properties of interspecific and intraspecific competition.
4. Learn how humans directly and indirectly influence ecosystem activities through competition with other species.

II. Introduction

All organisms, humans included, compete for resources, such as food, space, water, minerals, and sunlight. **Competition** may be defined as an interaction between individuals for shared or limited resources. Ecologists have long recognized that competition is an important ecological concept. It plays a major role in community structure, niche breadth and overlap, community succession, and evolution.

Competition can occur between individuals of the same species (***intra*specific competition**), or two different species (***inter*specific competition**). Either way, there is competition for resources that are limited. When organisms compete, they interact with each other in ways that affect their growth, reproduction, and/or survival. As a result of the competitive exclusion principle (introduced in Lab 5), competing individuals will adjust and coexist, relocate, or the less competitive individual will be driven to extinction. The final outcome depends on the degree of competition. For example, if the degree of competitive interaction is low, the niches of the species involved are different enough so that they can coexist. If competition is intense, only the species best adapted to that environment will survive. Since ecosystems are continually changing (albeit gradually), one species may be at an advantage under one situation, but at a disadvantage under another. When environmental conditions change, another species that is better suited may drive out the original "winner."

Since competition influences the organisms involved, and may take up more of their energy budget than they can afford, the best way for many organisms to survive competition is to avoid it. Animal species can change their feeding habits, time of day they forage, and/or places of feeding. When confronted with competition, the simplest response is for a species to move. Obviously, individual plants do not move, but as a population, plant distributions can change over time. Some plants secrete a substance that is harmful to other species, or even to other members of its own species. This strategy is known as **allelopathy.**

Over many generations, natural selection may favor genes that enable a species to exploit a part of the habitat (or niche) that its competitor does not occupy. Given a long enough time span, the previously similar species (or individuals) may evolve into entirely different species, thereby reducing competition as much as possible. In competition, species demonstrating the best adaptive traits ("stronger"), tend to increase in number, while less adaptive species ("weaker") become scarce. Stronger species tend to experience more intense intraspecific competition, because there are more of their own kind competing for common resources. Weaker species generally experience more intense interspecific competition, because they must compete with more successful species for resources. Weaker species, however, may improve their competitive abilities via natural selection, whereupon their numbers could increase. These oscillations will continue until relative stability is reached between competing species, or until some major disturbance (e.g., fires, floods, humans) creates different ecological conditions.

III. Relevance: Human–wildlife conflicts

As human populations increase, competition with wildlife for resources increases. This competition creates conflicts between humans and wildlife that must be resolved. A current conflict involves cattle ranchers and reintroduced endangered species like timber wolves and black-footed ferrets. Ranchers claim that wolves kill their livestock. Ranchers also kill prairie dogs, the main food of the ferrets, because livestock injure themselves falling into prairie dog burrows. This action has led to reductions in ferret numbers. Another conflict occurs between farmers in Africa and India, whose crops and livestock are eaten by native wildlife. How can these conflicts be resolved? How do we balance the needs of people against the needs of wildlife? Do you think there can be coexistence without conflict?

IV. Activity

In this lab, we examine competition among foraging birds. Flocking species, such as sparrows, finches, and blackbirds, often congregate at a common food source. During this time, both intra- and interspecific competition occurs. This competition often takes place in the form of fights, chases, displacements, and threat displays. The influence of competition can be seen by the fact that different species have different bill morphologies, and therefore different diets. Different species also feed at different heights, different distances from protective cover, and/or different times of day. These differences, termed *resource partitioning,* allow the species to occupy different niches, thereby reducing competition. During the laboratory period, you will be observing these interactions to measure competition.

V. Procedure

Materials

    bird feeders (simple platform feeders will do)

    sunflower seeds

    millet seeds

    field guide to local birds (if available)

    binoculars (helpful, but not essential)

    stopwatches

    data sheets

Method

In Class:

1. The instructor will have set up two bird feeders where you will make your observations. There must be three easily identifiable bird species located at these feeders. The two feeders will be set 1.5 m off the ground. One feeder will be filled with sunflower seeds, and also have sunflower seeds scattered on the ground beneath the feeder. The other feeder will be filled with millet seeds, and have millet scattered beneath the feeder. Approximately equal amounts of seed should be placed in each feeder, and on the ground.
2. The instructor will describe to the class the three common species that appear to be competing for the same resources in this situation. Only these three species will be used in this lab. Record the three species identified, and their identifying letter, on Data Sheet 6.1, section A.
3. Divide the class into at least three pairs (one group for each species) of students. Each group will be assigned a species to observe.

In the Field:

4. The groups should place themselves far enough away from the feeders so that they do not disturb the birds feeding there, but close enough to observe interactions. If available, binoculars will help students make observations from a distance. Each member of the different groups should choose a variable(s) to measure. The variables to measure are:
   a. Seed choice—choose an individual of the species for which your group is responsible and record the amount of time spent at the different feeders (regardless of whether the individual feeds on the feeder or the ground) for 1 minute. After the 1-minute period, select another individual and record the same information. Do this for 10 individuals.
   b. Foraging height—follow the procedure described above, except this time record the amount of time spent on the ground or on the feeder, irrespective of which feeder is being used.

Back in Class:

5. In the classroom, compile your data and calculate the averages on Data Sheets 6.1 and 6.2.
6. Answer the questions in Section VI.

Data Sheet 6.1

Name _____

Section _____

Date _____

A. Identify the three major species in your community:

    Species a _____

    Species b _____

    Species c _____

B. Which species will you be observing? _____

C. Fill in the columns in the table below with the length of time each of 10 individuals of your species spends at the feeder (sunflower or millet). Get the information for the other two species (10 individuals each) from the other groups back in class.

Obs.	Species a			Species b			Species c	
	Time spent eating sunflower seeds	Time spent eating millet		Time spent eating sunflower seeds	Time spent eating millet		Time spent eating sunflower seeds	Time spent eating millet
1								
2								
3								
4								
5								
6								
7								
8								
9								
10								
Total								
Ave								

D. To calculate the average for each group, divide each column total by 10.

Data Sheet 6.2

Name _____

Section _____

Date _____

E. Fill in the columns in the table below with the length of time each of 10 individual birds of your species spends on the ground or at the feeder (regardless of the feeder used).

Obs.	Species a			Species b			Species c	
	Time Time spent feeding at feeder	spent feeding on ground		Time Time spent feeding at feeder	spent feeding on ground		Time Time spent feeding at feeder	spent feeding on ground
1								
2								
3								
4								
5								
6								
7								
8								
9								
10								
Total								
Ave								

F. To calculate the average for each group, divide each column total by 10.

VI. Questions

1. Which type of competition (intraspecific or interspecific) seems to be the most intense? Explain.

2. How does the size of a species influence its proximity to its choice of diet and/or feeding height?

3. What other factors might account for the differences you saw?

4. In what ways has natural selection affected the morphology (body and bill shape) and/or behavior of the species you observed?

5. How do the basic features of competition influence such ecological processes as community structure and function, niche breadth and overlap, and evolution?

6. Why should humans be interested in studying competition between species?

EXERCISE 7

Is Your Campus Friendly to Wildlife?

I. Objectives

The student will determine whether the college campus provides necessary resources for the survival of wildlife. In this exercise, the student will:

1. Understand the basic needs of wildlife for food, water, and shelter.
2. Inventory your campus to determine its suitability for wildlife.
3. Develop plans to improve the wildlife habitat on your campus.

II. Introduction

As the human population increases, more people move into cities, suburbs, and towns. This process of urbanization results in major environmental changes, with native habitat being replaced by massive amounts of concrete, non-native ("exotic") vegetation, and building structures. Each of these components of an urban area pose hazards to many species of wildlife. The requirements for survival of many species—food, water, and shelter—will no longer be available. Animals must try to survive in an oftentimes dangerous environment. For example, roads pose extreme hazards to slow-moving amphibians (i.e., frogs), reptiles (turtles), and mammals (opossums). Birds must contend with new predators, such as cats, which kill thousands of birds each year. Cats are major predators, also, of other animals, such as squirrels. In fact, one of the best things a person can do to protect urban (and suburban) wildlife is to spay or neuter their cats, and keep them permanently indoors. The cat will get used to being an inside-only cat, and (besides not being able to hunt wildlife) will be much safer, since cars kill many cats, and sometimes neighbors poison trespassing animals. Birds also frequently fly into skyscrapers. In essence, they see reflected sky and fly right into the building, resulting in death or injury to the bird.

The enormous amounts of concrete involved in building urban areas decreases the availability of vegetation, which is essential for cover, and as a food source. Streams are often filled in, or too polluted to be of any use to wildlife. Exotic species, such as starlings and house sparrows, are good competitors, and can out-compete native species for the remaining shelter and nesting spaces. Other animals, such as rats, can pose major health risks. Unfortunately, it seems the animals that are best able to live alongside humans are those we consider "pests."

The structure of an urban area is also drastically changed. Instead of a diverse habitat, where many species exist, urban areas—including many college campuses—are remarkably homogeneous. That is, they consist of grass that is kept mowed and fertilized and has pesticides applied, with some trimmed bushes and trees planted around buildings. Some people maintain their suburban lawns with large amounts of pesticides, which can be poisonous to wildlife. The few green areas (e.g., parks) tend to be used by so many humans that many species of animals cannot live there, or reproduce successfully.

The good news is that many cities are beginning to include "green spaces" in their planning processes. City planners are leaving open spaces in new housing developments, and plan their new buildings around existing water bodies or streams. Suburban residents, too, can learn how to plant native vegetation, and design their backyards with wildlife in mind. By planting appropriate vegetation, and leaving "wildlife friendly" areas, even the smallest backyards can become animal havens. Many progressive architects are planning their buildings to minimize danger to flying birds, simply by putting netting or barriers against their windows, or tilting the windows to reflect the ground, instead of the sky. This consideration for wildlife, which increases the aesthetic value of property, in addition to providing essential resources for other species, takes some planning and forethought, but the results are well worth the effort.

III. Relevance: Animals Need a Good Home, Too

Urban areas tend to be extremely simplified ecosystems. A result of this ecosystem simplification is that only a few animal species will be present, such as house sparrows, grackles, starlings, pigeons, and feral (wild) cats. These species are able to eke out a living by making use of the resources found in cities and towns. Other species, however, cannot find the resources needed for survival and reproduction, so they leave the area. Urban areas, however, are not doomed to containing only these species. If the diversity of a habitat increases, so does the diversity of species. The good news is that urban areas such as college campuses can be designed, or redesigned, with diversity of both vegetation and animals in mind. Areas that are left "ungroomed" can provide suitable habitat for animals, since many types of bushes, small trees, and grasses grow in small plots of land. Not only does this save money (since people don't have to be hired to continuously mow the grass, for example), but it provides an interesting landscape, compared to the monotonous simplicity of lawns.

Many college campuses now plant native vegetation, and provide other necessities for wildlife survival. Environmental organizations on college campuses pitch in to pick up garbage, plant native vegetation, and clean up streams and ponds on the campus. Courses are now taught that emphasize natural forms of insect pest control (instead of pesticides), and the species of vegetation best suited for attracting and feeding wildlife. With a little bit of extra effort in the short term, urban areas—including college campuses—have seen a great increase in animal diversity and sustainability of the ecosystem for the long term.

IV. Activity

You will take an inventory of your campus to determine whether it supplies various wildlife species with the necessary food, cover, and water resources animals need to survive in that area. You will also identify the various species that inhabit the campus, and determine whether they are native or exotic species, and how they may have gotten to campus. Before you go out into the field, ask yourself what animals you remember seeing, and try to identify the resources each species needs for survival (e.g., what food do they eat? where do they spend their time? what predators may be present that will eat them?). You may also want to interview someone from the campus about whether an official policy exists towards attracting, or repelling, wildlife. Finally, how are you going to decide which species are "desirable," and which ones are "nuisance" species?

V. Procedure

Many kinds of wild animals do very well in urban settings as long as they are able to satisfy their basic needs for food, water, and cover to escape from predators, and the elements. You will survey your campus and make a map of the food and water sources and sheltered areas that could be used as cover. Look especially for the following kinds of situations:

1. Permanent water—ponds, swamps, bogs, lakes, creeks, irrigation ditches, and so on.
2. Food sources—fruits and seeds of trees and shrubs, grass and weed seeds, bird feeders, wildflowers in bloom, garbage, greenhouse refuse, gardens.
3. Cover—thickets of dense shrubs, open weed fields, hedgerows or fence lines, ditch banks, ravines, rock piles, landscape timbers, birdhouses, hollow trees, animal burrows, brush piles, landscape refuse, compost piles.

Method

1. Organize the class into work groups.
2. Survey the campus for sources of water, food, and cover for wildlife.
3. Develop a map showing where current resources are located.
4. Develop a plan for improving the wildlife habitat on your campus (Data Sheet 7.1).

Data Sheet 7.1:
Is Your Campus Friendly to Wildlife?

Name _____

Section _____

Date _____

A. Draw a map of your campus, showing where current wildlife resources are located.

B. Develop a plan to improve the quality of your campus for wildlife. Be particularly sensitive to providing ideas that do not require a high degree of maintenance or require substantial expenditure of money. For example, building and placing birdhouses around campus or allowing a field to grow into weeds would be preferable to suggesting that a pond be built or special food plots be planted every year. Obtain permission from the building and grounds department and implement your plan.

C. Describe elements of your plan regarding water, food, and cover in the table below:

Water	Food sources	Cover

VI. Questions

1. What permanent water sources are available for wildlife? Do these sources seem to be of good, or poor, quality? What implications does poor water quality have for wildlife?

2. What food resources are available? Are they from native plants, or introduced vegetation? Do humans play any part in feeding wildlife on your campus?

3. Is your campus heavily mowed? Are any parts left "wild," without use of fertilizers, pesticides, or mowing?

4. What does the campus sentiment seem to be towards wildlife? Do people complain about birds present?

5. Are there any potential trouble spots associated with wildlife on your campus? How can you minimize conflicts between wildlife and humans?

6. Describe some activities that you can do to improve the conditions for wildlife on your campus. What is the one major obstacle toward implementing your plan?

Field Trip Suggestions

1. Visit local habitats—grassland, desert, forest, or other locally available habitats.
   Collect five plants that you can take back to the lab to identify.
   Collect five invertebrate animals. Record in detail where each was found. Upon returning to the laboratory, identify the animals.
2. Visit various aquatic systems, such as a stream, a lake, a bog, an irrigation canal, or another locally available aquatic system.
   Collect five plants that grow in or adjacent to the water that you can take back to the lab to identify.
   Collect five invertebrate animals. Record in detail where each was found. Upon returning to the laboratory, identify the animals.
3. Visit various controlled ecosystems, such as a sewage treatment plant, an agricultural field, a municipal park, a forestry plantation, or another locally available area. Describe five ways in which each controlled ecosystem is different from a similar natural one.
   Describe five organisms that are aided by the human control of ecosystems and five that are harmed.

Alternative Learning Activities

1. Track an animal in the mud or snow. Identify the animal and try to determine what it was doing from the tracks you find.
2. Place some soil in a broth made by boiling hay and observe the changes in protozoa present over a two-week period.
3. Describe the stages of succession you can observe along the edge of a pond, in an abandoned field, or in a sidewalk crack.
4. Photograph several stages of succession you can observe in your neighborhood.
5. Participate in a local habitat modification project aimed at increasing the numbers of certain kinds of organisms.

Population Growth

Understanding nature and the human impacts on it are an important first step in understanding the various environmental problems and issues surrounding human existence on earth. One extremely important issue, and one that many feel is a major contributor to unsustainability, is human population growth. There are currently about 6 BILLION humans on earth, more than at any time in history. Millions of additional humans are added to the population each year, most in the poorest countries. More humans on the planet will require more food, shelter, education, health care, jobs, and space. Despite the fact that growth rates are dropping in several parts of the world, the sheer number of humans currently reproducing guarantees that billions more humans will be added to the planet within the next 40 years. These increased numbers of humans will put further pressure on the environment, causing more natural disruptions, more species to go extinct, more habitat destruction, and increased losses of natural ecosystems. It is vital that we learn not only about human population growth, but also how to control it, within cultural contexts and with sensitivity. Humans must be willing to make some hard choices regarding their population growth in order to avert disastrous consequences in the future.

Part 2 introduces the basic concepts of population growth. Exercise 8 uses fruit flies to estimate population growth rates, given the initial and final numbers of flies. Exercise 9 shows how human survival has increased during the past decades, and exercise 10 uses a mock United Nations debate to focus on methods to achieve a zero population growth rate for humans.

I. Objectives

Population dynamics addresses change in population size, composition, and distribution. After completing this exercise, the student will be able to:

1. Understand major concepts of population dynamics and the relationship to human populations and the environment.
2. Understand the concepts of exponential growth and logistic growth in populations.
3. Examine density-dependent population growth.
4. Gain an awareness of the extent to which population is a relevant concept.

II. Introduction

One of the basic units of study in ecology is the **population,** a group of individuals of the same species occupying the same area at the same time. **Population dynamics** refers to the changes that occur in the number of individuals in a population. Understanding population growth requires understanding the factors affecting that growth. These factors include reproductive rate, resource availability, and competition. Population growth is regulated by four factors: (1) **birthrate,** (2) **death rate,** (3) **immigration,** and (4) **emigration.** The balance between the number of individuals being added to the population and the number of individuals being removed from it determines whether the population grows or declines. In a closed population (no immigration and emigration, only births and deaths) with *unlimited* resources, populations could grow without any limits. If you were to graph population growth under these conditions, it would resemble a J-shaped graph (figure 8.1). This type of growth is called **exponential growth.**

In reality, however, biotic and abiotic factors (food, space, competition, etc.) limit population growth. The maximum number of individuals an area can support is the environment's **carrying capacity.** When a population exceeds the carrying capacity, the environment cannot support all the individuals. In response to this unsustainable population size, the death rate increases, birthrate decreases, and/or individuals leave the population (emigrate). Once the population falls below the carrying capacity, growth resumes. This type of population growth is called **logistic growth,** and results in an S-shaped, or sigmoid, graph (figure 8.2).

There are two categories of factors that limit population growth: **density-dependent** and **density-independent.** Density-dependent factors are those that affect populations in direct proportion to population size, becoming more severe as the population grows. For example, as populations grow, food supplies and space decrease, while predation, competition, and disease increase. Density-independent factors affect populations regardless of population size. For example, a hurricane that strikes a population is going to kill individuals, whether the population size is 10 or 10,000. Often in the real world, density-dependent and independent factors interact on populations.

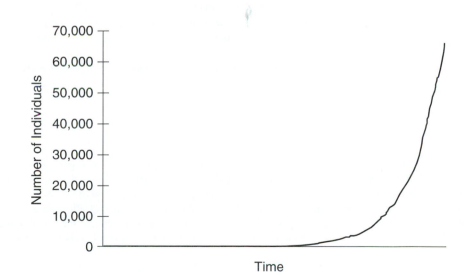

Figure 8.1 Exponential population growth

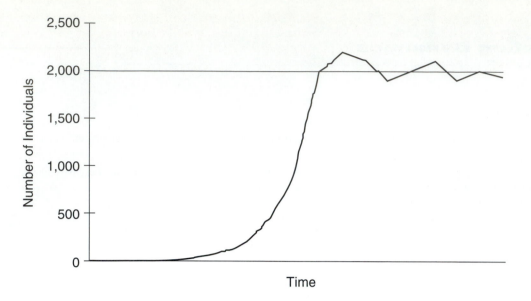

Figure 8.2   Logistic population growth

III. Relevance: How Many People Can the Earth Support?

There is probably no other subject that has a greater bearing upon the future for a sustainable environment and quality of life on earth than human population. Prior to the 1800s, some environmental degradation was occurring. However, with some exceptions, the impacts of wood burning, deforestation, mining, and raw sewage disposal, etc. were localized because human populations were still relatively low. Following the industrial revolution, advances in agricultural technology, medical science, and public health awareness, the human population (and impacts from their activities) grew exponentially.

Today, almost 6 billion people exist on earth. If it took you one minute to read the discussion on populations to this point, about 180 more people were added to the world's population as you read.

By the year 2000, an estimated 6 billion humans will exist on Earth, and by 2020, 9 to 10 billion humans are expected to inhabit the planet. Do you think the world is overpopulated? How about the United States? What environmental effects result from overpopulation in Third World countries, and the United States? How does consumption tie in with population in developed countries, such as the United States? Does overpopulation in other countries encourage migration (both legal and illegal) into the United States? Are there any problems (social, political, environmental) associated with migration? What are some social costs (and benefits) to having a large population?

It is important to note that in fact there *is* a limit to the number of people the Earth's resources will support on a sustainable basis, despite advances in technology. There may be room on earth for 10, 20, 30, or 50 billion people. However, the impact of feeding, clothing, providing transportation for sanitation, medical needs, cemetery space, power and utility demands, etc. would render *the quality of life* unacceptable in today's terms.

Dr. Paul R. Ehrlich, well known for his books on overpopulation, suggests that population growth causes a *disproportionately* negative impact on the environment. Ehrlich says that problems relating to population size and growth, resource use and depletion, and environmental deterioration *must all be considered together and on a global basis.*

IV. Activity

Fruit flies have been used in scientific studies for many years, because they reproduce rapidly, have short generation times, and are cheap and easy to care for. This lab uses fruit flies to observe the consequences of rapid, unchecked population growth. You will observe colonies of fruit flies that had different initial numbers of flies, and count the number of offspring produced at the end of two weeks.

Once the fruit flies have been sedated, they can be easily divided by sex and counted. Sex of fruit flies can be determined using the following information: female fruit flies are larger, and have pointed abdomens, with stripes extending to the tip of their abdomen; male fruit flies are smaller, with rounded abdomens and the last third of the abdomen is solid black, not striped. Your instructor will also help you distinguish between the sexes.

You will use the formula below to calculate the growth rates and doubling times for each colony. The following formula may look complicated, but if the directions are followed, you will find how easily it can be manipulated.

$$\mathbf{p_f = p_i * e^{rt}}$$

where: p_f = final population
p_i = initial population
e = 2.7183
r = rate of growth
t = time

This equation states that the final population size can be calculated by multiplying the initial population's size by some number that takes into account both growth rate and time. Any one of the above variables may be solved for; however, this lab solves for **r,** the rate of population growth, because we will already know all the other variables. The following steps explain the way to calculate **r.**

To simplify things, let x = rt. The formula then becomes:

$$p_f = p_i * e^x$$

When solving for e^x by dividing both sides by p_i, we get:

$$e^x = p_f/p_i$$

We can then divide the final population (p_f) by the initial population (p_i) and get a number which is e^x ["e raised to the x power"]. Now, here comes the hardest part: we must transform e^x into x. Each e^x has a unique value of x. Calculators often have a button "$\ln^x$" that will provide the x value once the e^x value is entered. [For example, if e^x equals 38.5, enter 38.5 in the calculator, press the $\ln^x$ button, and x = 3.65.] If your calculator does not have this function, the value of x can be located by using the graphs included in the Appendix by simply locating the e^x value you calculated, and looking at the related x value.

Since x = rt we now know x and t (the time the colonies had to grow), so we can solve for **r** (x = rt, so r = x/t). We can then transform it to a percent (since growth rate is always a percent) by multiplying the answer by 100:

$$r = x/t * 100$$

We have now calculated the growth rate (r) of that population.

Another powerful tool in studying demography is determining a population's **doubling time.** The doubling time indicates how fast a species' population is growing by determining the time it takes for that population to double in size.

We can determine the time it will take for that population to double in size by using the formula:

dt = 70/r, where dt = doubling time and r = growth rate (from the above formula)
[for example, if r = 2.1, then dt = 33.3 years (70/2.1)]

Once all the information is calculated, you will be able to compare the growth rates of the colonies, and draw conclusions about population growth and its effects upon the environment and other organisms. For example, are any of the colonies experiencing massive die-offs? If so, they might have surpassed the bottles' carrying capacity. What is the status of the resources (food, water, clean air, waste products)? What do you think will happen if the colonies are left for two more weeks? While humans certainly are not fruit flies, this experiment gives a broad perspective of what happens when a population approaches or exceeds the capacity of the resource base to handle that population. What do you think the future holds for humans if our population growth remains unchecked?

V. Procedure

Materials

5 labeled vials of live fruit flies

a commercially available fly anesthetic

cotton balls

dissecting microscopes or magnifying lenses (optional)

white sheets of paper

paint brushes

calculators

data sheets

Method

1. Divide the class into 5 groups. Each group will obtain one labeled vial of fruit flies that was prepared two weeks earlier. Read the label to record the initial (beginning) number of flies, on Data Sheet 8.1 (p_i).
2. Put a few drops of anesthetic onto a cotton ball or wand. **GENTLY** tap the edge of the vial on the desk to move the flies away from the top. Move the vial's plug slightly away from one side of the vial and **quickly** put the wand into the vial (or cover the top of the vial with the cotton ball). From this point on, *leave the vial on its side,* and leave the anesthetic in **only** until the flies stop moving (about 1–2 minutes). Remove the anesthetic and dump the flies onto the white sheet of paper (white paper makes flies easier to see).
3. Carefully divide the flies into female and male groups. Using the dissecting scope (or magnifying glass) and brush, count the flies of each sex and record these numbers on Data Sheet 8.1.
4. Count the dead flies in the vial and on the paper. Record this number on Data Sheet 8.1. [Dead flies will have their wings spread at a 90-degree angle away from their bodies; sleeping flies will have their wings folded close to their bodies.]

5. Carefully return the flies to their vials, leave the vials on their side (so the flies don't get stuck in the food), and return them to your instructor.
6. Put your observations on the board. Record the information from the other groups on Data Sheet 8.1, section B. Do the calculations on Data Sheet 8.2, and answer the questions in Section VI.

Pitfalls to Watch For:

1. To the instructor: Be sure to start fruit fly colonies two weeks before the lab.
2. Label each vial with the number of flies originally put into each vial (2 = 1 male/1 female; 4 = 2 males/2 females; 8 = 4 males/4 females; 16 = 8 males/8 females; 32 = 16 males/16 females).
3. Once the anesthetic is placed in or over the vial, LEAVE THE VIAL ON ITS SIDE, OR THE SLEEPING FLIES WILL FALL INTO THE FOOD AND WILL NOT BE ABLE TO BE COUNTED.
4. If dissecting scopes or magnifying glasses are not available, the flies can easily be distinguished by the naked eye.
5. Once the lab is over, the flies can be donated to the genetics department, or let loose.

Sample Calculations:

To calculate the growth rate (r) and doubling time (dt) of a population, when your results gave you the following information:

$$p_i = 20 \text{ flies (initial population of flies put into the vial)}$$
$$t = 14 \text{ days}$$
$$p_f = 81 \text{ flies you counted (total number of males and females)}$$

Using the formula, calculate r:

$$p_f = p_i * e^{rt}$$
$$81 = 20 * e^{rt}$$

remember x = rt, so:

$$81 = 20 * e^x$$
$$e^x = 81/20 = 4.05$$

Using your calculator, enter "4.05" then push the ln$x$ button to get:

$$e^x = 4.05, \text{ so } x = 1.40$$

Since x = rt,

$$r = x/t = 1.40/14 = 0.10$$

To get r as a percent, simply multiply by 100:

$$0.10 * 100 = 10.0\%, \text{ so the growth rate of that population is 10\%, and the doubling time is:}$$
$$dt = 70/r = 70/10 = 7 \text{ days.}$$

Data Sheet 8.1

Name _____

Section _____

Date _____

A. Record the following and put your information on the board:

   1. Initial number of flies: _____

   2. Number of males: _____

   3. Number of females: _____

   4. Total number of flies: _____

   5. Number of dead flies: _____

B. Using information from the board, complete the table:

Initial number of flies (p_i)	Number of males	Number of females	Total number of flies (p_f)
2			
4			
8			
16			
32			

Data Sheet 8.2

Name _____

Section _____

Date _____

C. Complete the following table by following the instructions, below:

Initial population p_i	Final population p_f	e^x	x	time (t)	Population growth rate (%) (r)	Doubling time (days)
2				14		
4				14		
8				14		
16				14		
32				14		

1. To determine the growth rate of each population:
    a. Record the final population (p_f) of each vial in column 2.
    b. To find each e^x, divide the final population by the initial population and record this number in the e^x column

$$e^x = p_f/p_i$$

    c. To find x, use your calculator (as described previously), or use the graphs provided in the Appendix. Record this value in the **x** column.
    d. Notice that **t** = 14 days.
    e. Remembering that x = rt, solve for **r**, and record in the appropriate column:

$$r (\%) = (x/t) * 100$$

D. To find the doubling time (**dt**), divide 70 by r (in %), and record it in the last column:

$$dt = 70/r (\%)$$

(*Note:* use the **r** value from step d; **do not** change it into a decimal.)

VI. Questions

1. Was the population growth rate (r) equal between the colonies? Explain any differences in the growth rate.

2. In your opinion, which, if any, of the colonies approached that vial's carrying capacity? Support your reasoning.

3. Why did we use fruit flies to illustrate rapid population growth?

4. Are there any human populations that approach the growth rate of the fruit flies?

5. Do the fruit fly vials accurately represent the earth's limited resources (i.e., food, fresh air, waste management, etc.)?

6. How does human population growth differ from that of the fruit flies?

Human Population—Changes in Survival

I. Objectives

Throughout the world, humans are living longer. After this lab, the student will be able to:

1. Understand differences in human mortality and survivorship between past and modern times.
2. Understand how changes in human mortality and survivorship have influenced population growth.

II. Introduction

The survival rate of humans in North America has increased significantly in the past one hundred years. Improved nutrition, preventive medicine, lifestyle changes, and new technologies are a few of the reasons for this improved life expectancy. Increasing life expectancies have had an impact on population growth rates. Put simply, there are more of us and we are living longer. In Rome during the first to fourth centuries, life expectancy was about 22 years at birth. Today life expectancy is approximately 75 years at birth in North America. Of particular note is the decline in infant and youth mortality in North America during the past one hundred years.

III. Relevance: Human Population

As more humans are living longer lives, they are placing additional and unique demands on the environment. With more babies surviving into adulthood, there are more potential parents, so that even if each adult female decides to have only two or three children, the sheer number of parents having children guarantees population growth. With the increased number of adults surviving to old age, environmental and social issues become increasingly important. The elderly must live in a clean environment, since they may be more vulnerable to pollution and environmental hazards. On the other hand, since retired adults often require social services and an income, many economists believe that an *increase* in population is necessary, to provide workers that can support those who have retired. But it is clear that an increased population places increased stress on an often overburdened environment. Thus, some type of balance between environmental and economic interests must be reached when dealing with population issues.

IV. Activity

You will obtain two sets of data, giving numbers of deaths in your community by age. The first set, representing vital statistics of your region in pre-1900 times, will be obtained from one or more cemeteries. (Some cemeteries have collected all the information from tombstones and have it recorded in their office. Local libraries often have this information as well.) The second set, representing current mortality figures, will be obtained from the obituary section of your local newspaper and the cemetery.

In order to prevent duplication of data, discuss in class what cemeteries are to be included in your study. Try to prevent the same cemetery (or section of a cemetery) from being counted more than once. Each student should record as many gravestones as necessary to give a total class sample of at least 100 males and 100 females. If you wish to do this exercise on an individual basis, you need not have such a large sample. Using the pre-1900 data sheet, record your entries for both males and females.

Obtain information on the deaths of 100 males and 100 females representing current deaths from the cemetery and obituary page of your local newspaper over the past five years.

To determine a survival curve, use the following method. Since you have data on 100 pre-1900 males, 100 pre-1900 females, 100 current males, and 100 current females, you can determine the number surviving to each age for each of the four groups by using the following technique. If you do not have 100 individuals in any of the four categories, you can convert your raw data into a percentage by dividing the number that died in each age group by the total number in the category.

V. Procedure

1. Visit local cemeteries, examine the information on tombstones, and record on Data Sheet 9.1 the age at death for males and females in your community who died before the year 1900.
2. Using the obituary page from your local newspaper and your local cemetery, record the age at death for males and females in your community who died during the past five years.
3. Enter the data on Data Sheet 9.2.
4. Plot the data on a graph (survival curve) on Data Sheet 9.3 for both and females for the two time periods. Use 4 different colored pencils for the two time periods and males and females.
5. Analyze the data and the reasons for change.

Example

Age at death (years)	Number that died	Percent surviving			
0	0	100 − 0	=	100	Plot the underlined numbers on the graph.
0–0.99	10	100 − 10	=	90	
1–4.99	15	90 − 15	=	75	
5–9.99	12	75 − 12	=	63	

Use these data to graph a survival curve for each of the four groups:

pre-1900 females

pre-1900 males

current females

current males

Use different-colored pens or pencils to record each of the sets of data on the graph on the data sheet.

Data Sheet 9.1:

Name _____

Section _____

Date _____

Cemetery (Pre-1900)

Age at death (years)	Male		Female	
	Number that died	Percent surviving	Number that died	Percent surviving
———	0	100	0	100
0–0.99				
1–4.99				
5–9.99				
10–14.99				
15–19.99				
20–24.99				
25–29.99				
30–34.99				
35–39.99				
40–44.99				
45–49.99				
50–54.99				
55–59.99				
60–64.99				
65–69.99				
70–74.99				
75–79.99				
80–84.99				
85–89.99				
90–94.99				
95–99.99				
100+				

Name _____

Section _____

Date _____

Cemetery and Obituaries (last 5 years)

Age at death (years)	Male		Female	
	Number that died	**Percent surviving**	**Number that died**	**Percent surviving**
——————	**0**	**100**	**0**	**100**
0–0.99				
1–4.99				
5–9.99				
10–14.99				
15–19.99				
20–24.99				
25–29.99				
30–34.99				
35–39.99				
40–44.99				
45–49.99				
50–54.99				
55–59.99				
60–64.99				
65–69.99				
70–74.99				
75–79.99				
80–84.99				
85–89.99				
90–94.99				
95–99.99				
100+				

Data Sheet 9.3

Name _____

Section _____

Date _____

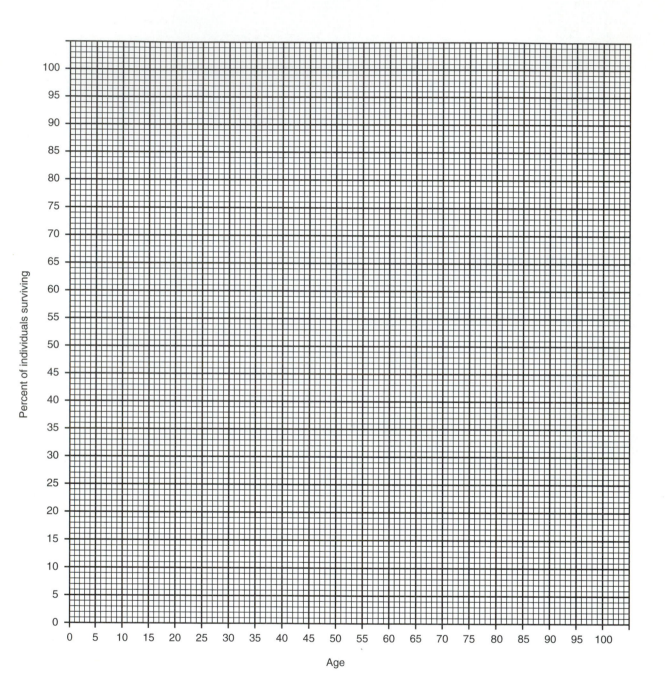

VI. Questions

1. What accounts for the difference in the four curves?

2. What do you think the curves would look like in the next century? What factors could influence the curves?

EXERCISE 10

Human Population Dynamics

I. Objectives

Human population dynamics addresses the growth of global human population now and in the near future. After completing this exercise, the student will be able to:

1. Calculate doubling time and growth rates for developed and developing countries, using the population growth formula.
2. Examine the effects of projected decreases in growth rates over time.
3. Investigate the impact of population growth rate on consumption of global resources and sustainability of human societies.
4. Conduct a mock United Nations debate, with each student representing a different country, to examine the difficulties inherent in reaching a consensus in achieving zero population growth.

II. Introduction

The growth rate of a species is the difference between the number of births and deaths in any given population. As discussed in the previous exercise, by knowing the growth rate, you can calculate a population's size for any time by using a simple mathematical formula. Human population growth can also be calculated given the growth rate, initial population, and time interval. Countries throughout the world have different growth rates, which will affect their population size, doubling time and impact upon the environment.

Generally speaking, developed countries tend to have much lower growth rates than developing countries. The overall growth rate (r) for developed nations is 0.4 percent and for developing nations 1.8 percent. Therefore, the doubling time (dt = 70/r) for developed nations is 175 years and for developing nations 39 years. The global human population of our planet is quickly approaching 6 billion. At the current population growth rates, in one year the populations of developed nations increased by 7.8 million people. During this same period, however, developing nations increased their population by 78 million people—adding ten times the number of people as developed countries.

Table 10.1 Growth Rates and Doubling Times for Various Countries

Region	Growth rate (%)	Doubling time (years)
World	1.4	50
Developed Countries	0.4	175
Developing Countries	1.8	39
Africa	2.5	28
Asia	1.6	44
North America	1.0	70
Latin America	1.7	41
Europe	0.2	350
Russia	0.3	233
Oceania	1.5	47

Source: United Nations' World Population Prospects, 1998 Revisions.

Both crude birth- and death rates affect the growth rate of a population. In order to bring the world's rapidly growing human population under control, we need to understand and manage the factors that control both birth- and death rates. To reach zero population growth (ZPG), the crude birthrate has to decrease faster than the crude death rate until the growth rate approaches zero percent. Although birthrates are dropping in many countries, the number of children added to the population each year is enormous. That is, even though many women are having fewer babies, there are millions of women having babies each year. Can countries achieve zero population growth? Yes they can, but it will take knowledge, commitment, and access to resources. Countries will take different amounts of time to achieve ZPG, depending on that country's growth rate, access to education (especially for women), and birth control. As the situation currently exists, the world's human population will continue to increase, until humans themselves decide to limit their growth rate.

III. Relevance: Population and Sustainability

The number of people on our planet intensifies the impact we have on natural systems. The human impact on our planet, both regionally and globally, is determined by the available resources in a region, and the stage of social development within that society. In developing societies, food is a basic necessity in high demand and low supply. Unfortunately, countries experiencing the highest population growth rates are also experiencing the worst food shortages. If Africa and Latin America, with the highest growth rates, cannot currently feed their present population, chances are they will not be able to feed future populations.

Developed nations are not immune from experiencing negative environmental impacts, however. These countries are characterized as having low population growth rates but are extremely energy-intensive societies. Developed societies critically impact a far greater range of global resources than developing countries. The long-range problem in achieving sustainability will be to manage the complicated relationships between populations and resources within the many different societies and cultures of the world. Presently, neither the developing nor developed countries are on paths that are sustainable. Each country has a unique set of problems in reaching sustainability; however, if we continue on the historical paths we are now pursuing, the result is nonsustainability for all. People are beginning to realize that it is not too late to reach sustainability—but we must begin now.

IV. Activity

In exercise 8, you studied population dynamics using fruit flies. Because they have a high reproductive rate, it is easy to demonstrate their population growth rate and calculate their doubling time. In this exercise, you will be given historical data of selected populations and various population growth rates. You will use the historical data and the basic formula from exercise 8 to calculate the final population size of humans in developed and developing countries.

Remember, the relationship between an initial (p_i) and final (p_f) population is:

$$p_f = p_i * e^{rt}$$

where: p_f = final population
p_i = initial population
e = a physical constant whose value is 2.7183
r = rate of growth
t = time

By defining $x = rt$, we can manipulate the formula to investigate several possibilities in human population growth. When you complete each section of the exercise, you will be able to: compare human population dynamics in various parts of the world; observe how zero population growth can be achieved in both more- and less-developed countries, and observe the impact a developed society has on energy resources.

You will also conduct a mock United Nations debate, with each student representing a different country, and, through discussion and debate, attempt to reach a consensus on whether all countries on Earth should achieve ZPG by 2025. (Details for the countries to be represented and facts to be researched are presented on the next two pages.)

V. Procedure

Materials
calculator

data sheets

Method
1. Divide the class into groups of two. Each group will do the following calculations.
2. In section A of Data Sheet 10.1, calculate the final population size for developing and developed nations, given the growth rate (r) and initial population size (p_i) for those countries. Use table 10.1 to obtain the data for developing and developed countries.
3. Calculate the population of both the developed and developing nations if the growth rate decreases from 0.6 percent by 0.1 percent increments every 10 years for the developed world (section C on Data Sheet 10.2); and from 2.0 percent by 0.4 percent increments every 10 years for the developing world (section D on Data Sheet 10.2). (If time is limited, you can do these calculations at home.)
4. Spend the rest of the class doing your UN debate. You must play the role of representing your country regarding its population size, growth rates, population control policies, etc. You may dress in the clothing representative of your country, and really "act the part" to increase enjoyment and participation in this debate.

SAMPLE CALCULATION:
To find the final population size of a country:

You are given the initial population size (p_i), growth rate (r) and doubling time (dt) of a country, so you can calculate its final population size (p_i). The method is very similar to calculations in exercise 8, with one major exception: the button you push on the calculator is NOT $\ln x$, but the e^x button (since you know r and t, and are trying to find the corresponding e^x).

$p_i = 4.7 \times 10^9$ (initial population of 4.7 billion people in developing countries)

$t = 39$ years (from table 10.1)

$r = 1.8\%$ (from table 10.1)

You can now calculate the final population size (p_f) of developing countries after 39 years:

$$p_f = p_i * e^{rt}$$

You must first change the rate of growth ($r = 1.8\%$) into a decimal by dividing by 100.

$$r = 1.8\% = 0.018$$
$$p_f = 4.7 \times 10^9 * e^{(0.018)(39)}$$

We'll put parentheses around the numbers to make things easier to see:

$$p_f = (4.7 \times 10^9) * e^{(0.018)(39)}$$
$$p_f = (4.7 \times 10^9) * e^{0.702}$$

Using your calculator, enter "0.702" then push the "INV" and "lnx" buttons to get:

$$e^{0.702} = 2.02$$
$$p_f = (4.7 \times 10^9) * 2.02 = 9.48 \times 10^9$$

That means that in 39 years, the population will be 9.5 billion people.

The United Nations Mock Debate

Objectives/Instructions:

1. You will work individually or in pairs to research your chosen country or interest group and participate in a class debate. This debate will take the form of a UN roundtable discussion on population issues.
2. The central concern of the UN meeting is Resolution 10445:

   "Be it resolved, that the world *must* achieve zero population growth (ZPG) by the year 2025."

   Remember that you will be representing your chosen group or country and that you must accurately present the information gathered from your research—this is not a forum for your personal views on the issue. In other words, unless you researched the United States, do not think like an American. This is an exercise in cultural appreciation, and as such represents an opportunity to examine different perspectives on a very controversial issue.
3. There are ten countries and two interest groups being represented. You will select one of these to research. Everybody is expected to participate in the research as well as the discussion. Your grade will be weighted equally on your written and oral presentation (50% each).
4. The debate may become heated, so be familiar with your material beforehand.
5. Your instructor may require you to write a report based upon the information you gathered. As with all reports, it *must* be typed and at least two different sources must be used. These can be found in books and catalogs at the library, on the Internet, or in on-line catalogs or microfiche in the Government Documents section of your library.

Country/Interest Groups:

United States, China, Russia, Mexico, Nigeria, India, Sweden, Brazil, Thailand, Saudi Arabia, Vatican City**, and Planned Parenthood**.

Report Format:

Present the required information (described below) in a chartlike format or a series of bullet statements on one page. On a second page, provide a one-page summation or general overview of your country, its rate and status of development, and an overall description—assume no prior knowledge on the part of your fellow "delegates." Based upon your findings, include your country's position on the UN resolution in particular, as well as population control in general. For example, do you believe this is a matter that should be addressed according to the rules of sovereign nations (autonomously) or as an international community? CLEARLY DEFINE AND EXPLAIN YOUR POSITION(S). Your country/group interests may or may not agree with what would be in the collective best interest of the other countries involved. Again, bear in mind that *you are acting as an advocate* for your country or group.

Be sure to provide information on what your country/group has done internationally and is willing to do in the future to comply with the resolution (or why it refuses to or cannot do so). *Hint:* You may want to petition to modify the resolution so it conforms more closely with your position, or to achieve consensus—save this for the day of the debate.

**For Planned Parenthood and Vatican City: Concentrate your efforts at defining your ethical and/or policy positions on family planning and population growth. BE AS SPECIFIC AS YOU CAN, using quotes, policy/mission statements, informational brochures, and literature, etc. Include information on specific roles your organization has played in the past and what efforts you would be willing to undertake in the future in order to advance your policies or beliefs. Specify the number of people/countries you represent worldwide (e.g., for the Vatican: how many Catholics there are, number of Catholic relief organizations, how many countries are predominantly Catholic in religious make-up; for Planned Parenthood: in how many countries do you have a chapter or equivalent representation?). Your written presentation should be at least two pages in length.

Required Information:

A. Demographic Data
   1. Population
   2. Population density (people per square km)
   3. Birthrate/death rate
   4. Population growth rate
   5. Total fertility rate (TFR) = the number of children a woman will give birth to in her lifetime
   6. Life expectancy (male and female)
   7. Infant mortality rate per 1000 (breakdown by male and female, if given)
   8. Child mortality rate per 1000 (as above)
   9. Maternal morbidity rate (urban vs. rural)
   10. Percent of population in urban vs. rural areas
   11. Predominant ethnic and religious groups, along with differences in ethnic group sizes
   12. Immigration/Emigration rates

B. Economic Data
   13. GNP per capita (in US $)
   14. Total GNP/GDP
   15. Agricultural base (self-sufficient vs. food importer)
   16. Primary exports/natural resources

C. Social/Political Data
   17. Education policy (compulsory? until what age/grade?)
   18. Literacy rate (male and female, if given)
   19. Health care (availability, physician/population ratio, subsidized, social vs. private system, nutritional status, etc.)
   20. Status of women (percent working, wages, voting, married vs. single, percent using contraception/family planning, etc.)
   21. Form of government/political history
   22. Country policy on population (support/discourage family planning efforts in own country, clinics and availability of family planning/prenatal care, cost of services, availability/legality of abortion, government funding for family planning programs, etc.)

Data Sheet 10.1

Name _____

Section _____

Date _____

Group Number: _____

A. From table 10.1, fill in the appropriate data for the growth rate and doubling time for developing and developed countries. Put your numbers in table 10.2, part (A).

The initial population size for developing countries, is $p_i = 4.7 \times 10^9$, and for developed countries, it is $p_i = 1.2 \times 10^9$. Use the exponential equation to determine the final population (p_f) for the developed and developing world.

$$p_f = p_i * e^x,$$ where $x = rt$ (remember to change the rate from percent to decimal)

B. Next, calculate the final population size (p_r) for the developing world if it had the *developing* world's high rate of growth and the *developed* world's long doubling time. That is, use the growth rate from the developing countries, but the doubling time from the developed countries. This will help you see the enormous population size of the developing countries in 117 years, if the growth rate does not decrease.

Table 10.2

	Region	r (%)	dt (years)	p_i ($\times 10^9$)	p_f ($\times 10^9$)
A	Developing world			4.7	
	Developed world			1.2	
B	Developing world		**	4.7	

**Same time as developed world dt

Data Sheet 10.2

Name _____

Section _____

Date _____

Group Number: _____

Use the same exponential formula to calculate what would happen to the population size if growth rates were gradually decreased every ten years. That is, you will do the same calculations as described earlier, but with TWO differences: (1) the final population of, say, generation 2 becomes the INITIAL population of generation 3, and the final population of generation 3 becomes the initial population of generation 4 and so on, and (2) for each generation, the growth rate (r) will decrease by 0.1 (developed nations) and 0.4 (developing nations). This decrease in r means that birthrates are getting closer to death rates, and hence the country is reaching ZPG.

C. Calculate the final population (p_f) for developed nations where (r) starts at 0.6 percent and decreases by 0.1 percent every ten years until (r) = 0.0 percent. Remember, the final population (p_f) becomes the initial population (p_i) for the next ten-year period.

Developed World

r (%)	t (years)	p_i ($\times 10^9$)	p_f ($\times 10^9$)
0.6	10	1.2	
0.5	10		
0.4	10		
0.3	10		
0.2	10		
0.1	10		
0.0	10		

D. Calculate the final population for developing nations where (r) starts at 2.0 percent and decreases by 0.4 percent every ten years until (r) = 0.0 percent. Remember, the final population (p_f) becomes the initial population for the next ten-year period.

Developing World

r (%)	t (years)	p_i ($\times 10^9$)	p_f ($\times 10^9$)
2.0	10	4.7	
1.6	10		
1.2	10		
0.8	10		
0.4	10		
0.2	10		
0.0	10		

VI. Questions

1. Considering the information derived in section B of Data Sheet 10.1, do you believe the developing world's population will reach this number in 175 years? What percent of the world's total population would live in the developing world if the growth rates were to continue as they are presently for the next 175 years? Do you think the developing countries could adequately deal with this many people?

2. Comparing the data derived on Data Sheets 10.1 and 10.2, what would the population of the developing world be after approximately 70 years if there was no decrease in the population growth rate compared to a decrease of 0.4 percent every ten years? (*Hint:* double your answer for the developing world in section A so that you have the population after 66 years.)

    Pop. of developing world = _____ with constant growth rate.
    Pop. of developing world = _____ with decreasing growth rate.

3. Which of these two population figures do you think is more realistic for the developing world in the year 2060? Do you think the lower of these two populations is sustainable on our planet?

4. What strategies may the developing countries start using right now to decrease the population growth rate?

5. What do you think is most important for the developed world and the developing world, respectively, to decrease the rate of population growth or to decrease the per capita energy consumption?

Field Trip Suggestions

1. Visit an orchard, cotton field, or vegetable farm. Determine which pesticides are used and what the purpose of each pesticide is. Figure out the cost of the pesticides used each year. Determine what special training is required for pesticide applicators. What conditions determine when pesticides are applied?

Alternative Learning Activities

1. Have a person from your local extension service office come to the class to discuss the pros and cons of pesticide use and the licensing requirements for pesticide applicators in your state.

2. Research the history of Planned Parenthood and the contribution of Margaret Sanger to the movement. Write a paper on the history of reproductive enlightenment.

3. Survey a community and ask questions about desirable family size and actual family size. Correlate your data with the age of the respondent.

4. Prepare a graph showing the population growth of your local urban community over the past twenty years. Select five changes (political, economic, sociological, housing, etc.) in the community, which appear to be related to population change and describe how they are related.

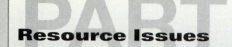

PART 3

Resource Issues

People living in the more-developed countries are not innocent of environmental destruction. Although the growth rates, and size, of the population may not be as high as in less-developed countries, developed areas use far more resources than their poorer neighbors. In fact, developing countries often correctly chastise people in Europe and the United States for using vast amounts of resources to live a luxurious lifestyle, while they struggle to merely survive. Inefficient use of natural resources (air, water, and soil) cause vast amounts of pollution. Despite laws designed to protect the quality of our air, water, and soil, there is still much work to be done to protect these resources, since our quality of life depends on maintaining the quality of our environment. Technology can help prevent, or clean up, pollution, but technology is expensive. The best way to maintain natural resources is by using the minimum of resources with maximum efficiency, and providing alternatives that decrease pollution to levels that are environmentally acceptable.

Part 3 introduces some basic concepts of natural resources and pollution. Exercises 11, 12, 13, and 14 describe various aspects of water: treatment of sewage, measuring pollution in the laboratory and in the field, and using biological indicators to determine whether the water is polluted. Exercise 15 examines air pollution, using laboratory and field measuring techniques. Exercise 16 examines various aspects of soil nutrients and characteristics.

I. Objectives

After completing the work associated with this exercise, the student will be able to:

1. Understand the infrastructure and processes involved with protecting and treating local water resources.
2. Tour the local drinking water treatment and wastewater treatment facilities and learn the steps utilized to treat the water before and after use by municipal and industrial users in order to protect the public health and the environment.
3. Gain an awareness of water quality and quantity problems in your own community, and an appreciation of water conservation on a global scale.

II. Introduction

Because of water's unique molecular properties, it is called the "universal solvent." Given enough time, there are few substances that will not dissolve in water: rocks, metals, wood, even human-made materials. Because of this characteristic, water becomes the transporter of most everything that it comes into contact with, whether that be pristine limestone formations in an aquifer thousands of feet below the surface, or with the garbage thrown into roadside ditches, or buried by the ton daily in many landfills.

Over the billions of years the earth has existed, a cycle of water evaporation, precipitation, infiltration (the seepage of water into the soil), runoff, uptake by plants, and eventually back to evaporation, has cleaned the water in a system called the **hydrologic cycle.** As long as the ability of this cycle to clean the Earth's water is not exceeded, the most precious of Earth's resources will remain clean. Unfortunately, about the beginning of the Industrial Revolution, signs of overload on this system began to show up.

Is water quantity a problem? Will we "run out of water," as some have asserted? From the standpoint of humans having clean unpolluted water, this could be possible. If there were 20 to 50 billion people on Earth using water at a rate and to the degree of abuse we in the United States are using it, then yes there might not be enough to go around.

Contrary to popular belief, however, we are not going to diminish the amount of water on earth. There is exactly the same amount of water on Earth today as there ever has been. No matter what we do to water, we can not reduce its *quantity.* We can only change its form (liquid to vapor or to solid, etc.), its quality (pollution), or its location at any one point in time (damming, diversion, pumping for irrigation, etc.). There is enough water that falls on the Earth each day for every man, woman, and child to have five gallons of water. It may not fall in Phoenix or Los Angeles, where so many want to live, but it falls. It is only a matter of distribution—in oceans, ice caps, groundwater, surface water (most fresh water is tied up in the ice caps and in groundwater). So what does this tell us about the real problem with water at our current population level? It is not quantity. The most immediate problem facing us is with the *quality* of water. That is, while the total amount of water may not change (although the ratio of ice, vapor, and water may), the quality of water may be decreased to such an extent that it will become unusable.

III. Relevance: There Is No "Away"

Americans often take their utilities (e.g., electric and water supplies) and services (e.g., garbage and sewage) for granted. They believe that things just "go away" when they throw their garbage out or flush their toilets. It is easy to become complacent with the high quality of potable water we have in the majority of the United States. Increased knowledge and an increasingly technical array of equipment is applied to monitor the quality of raw water supplies today to ensure that when you turn on the tap at your home, the water is clean. Additionally, the way we treat our wastewater to reduce its potential toxicity is just as important. When you flush your toilet, the waste does not magically disappear. Those wastes must be dealt with, lest they turn up in our drinking water. Again, knowledge and technology are helping to clean up what we flush down.

As populations increase, it is critical that citizens become aware of the role they play in helping to protect the water supply. It will not matter how much knowledge, how much equipment, or how much money is thrown at water-quality problems if populations do not act diligently and responsibly to protect this resource. To this end, you will be afforded an opportunity to see water protection and treatment in action in hopes of creating an awareness in you, which can help make a difference.

IV. Activity

The class will tour the local drinking water treatment facility (see figure 11.1) to learn what lengths the city goes to ensure that the public has clean water, as well as touring the wastewater treatment facility to learn what has to be done to the wastewater before it is released to the environment. This lab is preparation for the following lab on water quality, wherein we will examine specific water-quality parameters.

Before you go to the facility, think about the following questions, and be prepared to address them as you go through the tour.

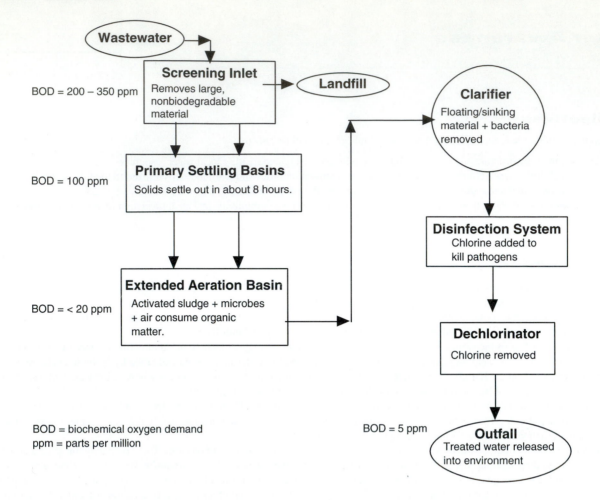

Figure 11.1   A typical wastewater treatment system requires 12 to 24 hours and treats 20 to 60 million gallons per day.

1. What is the quality of the water coming from the raw water source to the drinking water treatment facility and how does this level of quality compare with other city water sources (be specific)?
2. What are the treatment parameters with which plant operators must be concerned for the raw water (list at least five)?
3. What anthropogenic (human-made) activities are occurring upstream that influence the quality of the raw water?
4. What are the results and effects of these influences?
5. What measures do the operators have to take to allow for these influences?
6. What diseases and ailments would likely affect the population if there were no treatment of drinking water?
7. Why is chlorine added to the water at the drinking water treatment stage? Why is it added to the water prior to distribution to the public?
8. Why is it necessary to treat the wastewater coming from municipal and industrial effluents?
9. What is the major treatment process called that is used at the wastewater treatment facility to break down the organic solids in the sewage?
10. Where do these organisms originate?
11. Why is chlorine removed from the water prior to release to the receiving body of water? What problem does this step address? What is the receiving body of water?
12. What happens to the sludge that results from the sedimentation processes?

V. Procedure

Materials

You will need to wear appropriate clothing for the tour, and have your lab manual and questions, writing supplies, and a clipboard or other backing on which to draw and write.

Method

Before the field trip, write the questions down, and try to answer them. Also, before the trip, think of additional questions that have local importance, write them down, and ask them during the tour. During the tour, ask the guide questions you do not have answers to. Remember, it is up to you to ask the guide questions.

Data Sheet 11.1

Name _____

Section _____

Date _____

Location: _____

Person conducting the tour: _____

Draw a diagram, using boxes and arrows, showing the complete treatment process applied at the drinking water treatment facility, tracing the flow of water through the facility, including the chemicals added, where they are added, and treatment steps with labels for each segment (sedimentation, sand filtration, etc.) from the point the water is taken from its source to the point it is supplied to the pressure mains for public use. Use the back of this sheet also, if necessary.

Data Sheet 11.2

Name _____

Section _____

Date _____

Location: _____

Person conducting the tour: _____

Draw a diagram using boxes and arrows showing the complete treatment process applied at the wastewater treatment facility, tracing the flow of water through the facility, including the chemicals added, where they are added, and treatment steps, with labels for each segment (sedimentation, filtration, etc.), from the point the water is received at the screening inlet point to the point it is released to the environment. Use the back of this sheet also, if necessary.

VI. Questions

1. What was the most interesting part of your field trip?

2. Do you see any shortcomings in the treatment of the sewage that enters the facility?

3. What happens if toxic wastes are accidentally spilled into the wastewater stream? For example, what may happen to the biological organisms that digest the organic material?

4. Are you satisfied with the current treatment of drinking water and sewage? How could this treatment be improved?

Water Pollution

I. Objectives

After completing the work associated with this exercise, the student will be able to:

1. Understand the implications of water pollution.
2. Perform a basic experiment to determine the water quality of a given system based on biochemical oxygen demand (BOD).
3. Gain an awareness of water pollution problems in your own community.

II. Introduction

Water is the foundation of life on this planet. Without clean water we would soon perish. Unless we quickly change our attitude about water pollution and water conservation, that may actually happen. Even though 70 percent of the earth's surface is covered with water, only 3 percent is fresh water, and less than 1 percent is usable.

Water pollution is defined as the release of substances into the environment that ultimately enter our water, in quantities that are harmful to organisms in the water or that use the water. It is hard to determine what level of water pollution irreversibly damages the environment. In order to save this resource for future generations, we need to err on the cautious side today.

The most obvious source of water pollution, called **point source pollution,** is directly discharged into water. Examples may include pipes coming directly from factories, open ditches, or sewage discharge. A less obvious source of pollution, called **nonpoint source pollution,** comes from a large, diffuse area, which includes air and land. Examples include pesticide and fertilizer runoff from farmland, urban runoff from storm drains, and acid deposition from air pollution. Nonpoint source pollution is more difficult to detect and treat than point source pollution, but is the major source of water pollution in the United States.

Most water pollution can be divided between organic and inorganic pollutants. **Organic pollutants** are those that occur naturally, such as from sewage, agriculture, and food processing. Organic pollutants are usually not toxic themselves, but in large quantities they reduce the dissolved oxygen (DO) content of water. Organic pollutants may also result in bacterial growth that may present a human health hazard. Beaches near many large urban areas sometimes have to be closed because sewage runoff has led to bacterial buildup in the water. **Inorganic pollutants** are usually not biodegradable; they tend to stay in the system, and can become amplified through the food chain. Examples include acids, heavy metals, oil, phosphates, nitrates, and some pesticides.

Major environmental effects of water pollution are reduction of dissolved oxygen, elevated levels of toxins, increased turbidity, and elevation of water temperatures. Potential ecological consequences include elimination of species, altered photosynthetic activity, and changes in community structure and function. We can reduce the impact of water pollution by improved treatment of waste before discharge, careful design and operation of systems, developing new technology, and tighter regulatory controls.

There have been many specific tests designed to test water quality. Most of these tests are specific for one type of pollutant, and do not give a broad interpretation of water quality. In testing for organic pollution, a common method to determine general water quality is the biochemical oxygen demand (BOD) test.

III. Relevance: An Indirect Method of Measuring Quality

The BOD method indirectly analyzes organic content by measuring the amount of oxygen consumed by the bacteria and algae living in the contaminated water. Organic matter (such as sewage sludge, livestock runoff, and slaughterhouse waste) itself is not toxic in water. In low concentrations, it may even act as a fertilizer (it is, after all, simply organic material), but in larger concentrations it often produces toxic effects by encouraging the growth of bacteria and algae, which use up the oxygen in the water. As algae undergo metabolism, it uses the oxygen to run the process. While algae is part of the natural aquatic ecosystems, it may grow so rapidly that it causes the ecosystem to become unbalanced. The higher the organic content, the more algae grow and the higher the oxygen demand. Thus, as the oxygen demand increases, the oxygen content in the water decreases, and other organisms (e.g., fish species) that also require dissolved oxygen die. There have been cases where thousands of fish die at one time due to oxygen depletion. Another problem with water pollution is that animal and human wastes may contain disease-causing bacteria, such as cholera and typhoid. There are simple tests that can, again, indirectly measure the potential for such dangerous bacteria being present in the water, and, if these bacteria are present, more sophisticated tests can be run, and the water treated to kill the bacteria.

IV. Activity

Today's lab exercise measures how much oxygen is consumed through respiration and decomposition of microscopic organisms. Before we can measure the BOD of a water sample, we must know the dissolved oxygen (DO) content of the sample. The amount of DO in water affects natural biochemical processes in two ways: by limiting the amount of oxygen available for respiration, and by affecting the solubility of essential nutrients in the water. Four major factors that affect the dissolved oxygen content of water are: (1) diffusion at the air–water interface, (2) the photosynthetic activity of plants, (3) the bacterial breakdown of waste products, and (4) the respiratory activities of other organisms in the water. In addition to these four, wind currents will affect the distribution of dissolved oxygen in open water, and water temperature can affect how much oxygen the water can hold. Cold water can hold more oxygen than warmer water.

The decomposition of organic matter is the greatest factor in the depletion of available oxygen in a given body of water. As the organic content of water increases, there is a corresponding increase in bacterial activity and a decrease in the dissolved oxygen content. This can be illustrated by a graph known as an oxygen sag curve, which plots, using arbitrary units, the dissolved oxygen concentration against distance or time from the point of discharge of the organic pollution (figure 12.1).

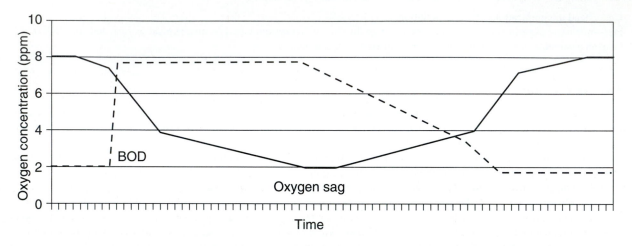

Figure 12.1   Oxygen sag curve

The dissolved oxygen content of the water is measured under saturated conditions (the water sample contains the maximum amount of oxygen it can hold) and again after incubating for five days in the dark at approximately 20° C. The difference between the two values is the amount used by organism decomposition and is known as the biochemical oxygen demand (BOD).

$$(1) \qquad BOD = DO_1 - DO_5$$

The dissolved oxygen content is measured using the BOD method. In this method, a treated sample of water is titrated with a standard solution using starch as an indicator. Every milliliter of *titrating solution* used indicates the presence of 0.02 mg of dissolved oxygen. Therefore:

$$(2) \qquad mg/{\leq}DO = ml\ PAO \times 0.02$$

BOD provides a fairly reliable standard with which to compare the relative amounts of organic matter in an aquatic system. The acceptable degree of BOD loading in a particular aquatic system varies widely, and can usually be determined by experience with similar systems. Remember, the cleaner an aquatic system is, the closer its BOD rating is to zero. The opposite is true for dissolved oxygen. An aquatic system is healthiest with a high amount of DO in the water. In this experiment, we will sample water from two different sites so we can compare them and determine if there is a difference, and if either of them are dangerously polluted. **Any sample with a BOD rating of 5.0 mg DO/L H_2O/5 days or greater should be further evaluated**.

Note: This lab requires measuring DO twice: at day one, and at day five. If this lab will consist of just one meeting, the instructor will have to measure the first DO five days prior to the lab's regular meeting. The students will then perform the lab procedures from step 4, simply getting the DO data of day one from the instructor. Alternatively, the lab instructor may want the class to come in on their regular lab day and perform the day one DO measurements, and then come in on their own to perform the day five DO measurements. Finally, if time permits, the instructor may make this a two-session course; the students would measure DO at day one, and day seven, with little effect on accuracy.

V. Procedure

Materials

Each work station needs:

2 300 ml BOD bottles w/ground glass stoppers

1 air outlet with rubber tubing

1 dissolved oxygen test kit

1 Erlenmeyer flask (250–500 ml)

Method

1. Obtain water samples from two places you want to test and compare organic pollution levels (i.e., rivers, lakes, ponds, groundwater, storm drains, tap water, etc.). Each pair of students should completely fill their BOD bottles with their assigned water samples. One set of bottles will be tested upon return to the lab. The second sample will be stored for 5 to 7 days.

2. Using the air outlet and rubber tubing, aerate the raw water samples for at least 10 minutes to saturate them with air. Insert a stopper into the bottle and make sure there are no air bubbles trapped in the neck. Cap one bottle with the plastic cap to prevent evaporation and air exchange. Place the bottle in the incubator at 20° C for 5 days. If an incubator is not available, simply place the bottles in a dark location (e.g., a cabinet) that will stay at room temperature.

3. Half of the class will test one water sample (e.g., river water), and the other half will test the other (e.g., tap water). After obtaining raw data, the groups will exchange data. Two students should work at each work station.

4. The test will be run using commercially available test kits. Use the following general instructions; or the instructions that came with the kit.

   a. The sample must first be "fixed" so that oxygen exchange will not occur during testing. To accomplish this, add the contents of one manganous sulfate packet and one alkaline-iodide-azide packet to the sample. Insert the stopper and invert the bottle until the powders have completely dissolved. The solution will change color (to orange-brown or white), and a precipitate will form. Let the precipitate settle, then add the contents of one sulfamic acid packet. Stopper and mix. The precipitate will disappear, leaving only a clear brownish solution. The sample is now fixed.

   b. Pour 100 ml of the fixed solution into an Erlenmeyer flask. This is your Trial 1 run.

   c. Insert a clean delivery tube into a 0.200 sodium thiosulfate cartridge and attach the cartridge to the titrator. Move the plunger on the titrator down until a few drops of sodium thiosulfate solution come out of the end of the delivery tube.

   d. Place the end of the delivery tube into the flask containing the water sample. Turn the delivery knob on the end of the titrator, swirling the flask every few turns. Continue until the solution in the flask turns pale yellow. The reaction can occur quickly, so do not turn the delivery knob too quickly. If the solution turns clear, too much sodium thiosulfate has been added, and you will have to start over with a new water sample.

   e. Remove the titrator and add 2 ml of starch indicator solution to the water sample. Swirl the flask to mix. The solution will turn dark blue.

   f. Continue titrating as before, and stop when the solution turns colorless. Record the number on the side of the titrator on the Data Sheet 12.1 table in section A,. This is the number of times you turned the delivery knob and represents the number of ml of PAO solution titrated. Multiply this number by 0.02. This number is the mg/L of dissolved oxygen in your sample.

   g. Repeat step 4 twice to obtain data for Trials 2 and 3.

5. Record the data obtained from the other half of the class.

6. After five days remove the first bottle from the incubator and repeat steps 4a–g. Calculate the BOD using equation (1).

7. Do calculations on Data Sheet 12.1, and calculate derived data for the other sample on Data Sheet 12.2.

8. Answer the questions in Section VI. Your instructor will inform you if a report is required for this lab.

Data Sheet 12.1

Group number: _____

Location: _____

A. Raw Data:

Day 1

Trial #	ml PAO used
1	
2	
3	
Average ml PAO used	

Day 5

Trial #	ml PAO used
1	
2	
3	
Average ml PAO used	

B. Derived Data:

DO_1 = average $\times$ 0.02 = _____ mg DO/L H_2O

DO_5 = average $\times$ 0.02 = _____ mg DO/L H_2O

BOD = $DO_1 - DO_5$ = _____ mg DO/L H_2O/5 days

Data Sheet 12.2

Name _____

Section _____

Date _____

Group number: _____

Location: _____

C. Raw Data:

Day 1

Trial #	ml PAO used
1	
2	
3	
Average ml PAO used	

Day 5

Trial #	ml PAO used
1	
2	
3	
Average ml PAO used	

D. Derived Data:

DO_1 = average $\times$ 0.02 = _____ mg DO/L H_2O

DO_5 = average $\times$ 0.02 = _____ mg DO/L H_2O

BOD = $DO_1 - DO_5$ = _____ mg DO/L H_2O/5 days

VI. Questions

1. Which of the factors that affect the oxygen content of water will add oxygen to the water, and which will reduce oxygen content?

2. Which of these factors were eliminated when you put your sample in a stoppered BOD bottle and placed it in a light-free incubator?

3. Which of the two test sites had the highest BOD rating? What would you conclude is polluting the water that made this test site have a higher rating?

4. Are either or both of your results dangerously high? If so, what can or should be done about the pollution?

5. Based on your experience with this lab, do you think water pollution is a problem in your area? In the nation? In the world?

EXERCISE 13

Stream Ecology

I. Objectives

After conducting this exercise that examines and compares various aspects of two streams, the student will be able to:

1. Compare several differences between polluted and unpolluted streams.
2. Learn techniques for evaluating the quality of water in a stream.

II. Introduction

The kinds of plants and animals that can live in a stream are determined by several factors, such as the following:

- Sediment that enters the stream from the land the stream drains
- Organic compounds that enter with sewage, falling leaves, and land runoff in the drainage basin
- Oxygen concentration of the water
- Temperature of the water
- Fertilizers that enter the water from farmland and lawns
- Toxic materials that enter from the air or from runoff
- Bacterial levels

This exercise compares two kinds of streams: one that is reasonably clean and gets most of its water from undisturbed areas, and one that is reasonably dirty and receives runoff from residential and agricultural areas. A field trip will be necessary in order to visit both types of streams.

III. Relevance: Streams in Danger

Many plants and animals depend upon a clean source of water for survival. Fish, aquatic vegetation, and invertebrates exist in rivers and streams, and are vulnerable to changes in physical (e.g., temperature, pH), biological (e.g., introduced predators, diseases), and chemical (e.g., oxygen, nitrogen) parameters. Unfortunately, many of the world's streams, rivers, and estuaries are the dumping grounds for pollution, both intentional and unintentional. Many developing nations do not have the facilities to dispose of their sewage properly, and simply dump their wastes into the nearest stream. Developed nations also cause significant damage to their rivers and waterways. Runoff from dairies and feedlots (called confined animal feeding operations or CAFOS) has polluted many thousands of river miles. Runoff from agricultural fields, as well as suburban housing areas, contains pesticides, fertilizers, and sediments. These substances end up in waterways and lead to increased nutrient levels, which decrease oxygen, light penetration, etc., and can affect survival of many different species.

IV. Activity

In this exercise, you will take a field trip to two local streams and compare their quality using physical, chemical, and biological characteristics of the streams. These abiotic and biotic parameters help determine which organisms can live in the stream. Students will test various aspects of each of the characteristics of the water by using simple, standard equipment that many professionals use. The **physical character** of the stream involves testing the temperature, pH, and total suspended matter in the water. The **chemical nature** of the stream can provide information on the amount of pollution present in the water, such as phosphate and nitrogen compounds, as well as the amount of oxygen dissolved in the water. Finally, the **biological nature** of the stream gives estimates of the organisms, both microscopic and macroscopic, that live in the water. Such organisms can provide information about the amount of pollution in the water.

Each group of students will collect different data, to save time and make the field work more efficient. By dividing into a maximum of seven groups, the class can gather lots of data, share the information, and draw conclusions about the two streams. If time or supplies are limited, you can select which parameters the class will measure, and ignore the other tests.

V. Procedure

Materials

In the Laboratory for General Use:

Physical parameters:

desiccator

water filter apparatus

balance

oven

Chemical parameters:

nothing required

Biological parameters; bottom organisms:

nothing required

Biological parameters; standard plate count:

sterile water

tryptone glucose extract agar

warm water bath

incubator for bacterial samples

Biological parameters; coliform bacteria:

lactose broth

incubator

most probable number table of coliform bacteria

Per Group:

Physical parameters:

thermometer

pH meter

1 liter flask with stopper

glass fiber filter disk

graduated cylinder

small stainless steel or aluminum dish

Chemical parameters:

oxygen measuring kit

total phosphate measuring kit

nitrogen measuring kit

Biological parameters, bottom organisms:

net for catching bottom-dwelling stream organisms

container with stopper to hold stream organisms

container with stopper to hold stream sediment

microscope

identification guide for benthic (bottom dwelling) organisms

Biological parameters; standard plate count:

sterile bottle with stopper

sterile flask for water blank

5 sterile pipettes

4 sterile petri dishes

Biological parameters; coliform bacteria:

sterile bottle with stopper

9 fermentation tubes containing lactose broth and a small inverted vial

4 sterile pipettes

Method

The class will be divided into groups, with each group having a specific task to accomplish. Each group will have equipment to use to gather data for its part of the exercise. Please follow the directions provided in this exercise or with the specialized equipment you will be using and record results on Data Sheet 13.1.

1. Physical Character of the Stream (Group I)
    A. Temperature
        Use a thermometer to take a temperature reading of the stream. This reading should be taken in a free-flowing portion of the stream. In some situations, you may need to attach the thermometer to a pole or wade into the stream. Record results.
    B. pH
        The pH of the stream can be measured using a portable pH meter. Take measurements in a free-flowing portion of the stream. Record results.
    C. Total suspended matter
        Use a standard container, such as a 1-liter flask, to capture water from a free-flowing portion of the stream. This can be stoppered and returned to the lab to determine the amount of suspended solids in the following manner:
        1. Obtain a glass fiber filter disk from the desiccator and weigh it very carefully (to the milligram). Record the weight.
        2. Assemble the filter apparatus as directed by your instructor.
        3. Thoroughly mix the water in your sample to resuspend any solids that might have settled to the bottom.
        4. Filter 100 ml of the sample.
        5. Carefully place the filter on a stainless or aluminum dish and dry in an oven at 103°–105° C for one hour.
        6. At the end of one hour, cool the filter in a desiccator. Then weigh it and record the weight.
        7. Repeat the drying and weighing cycle until the weight doesn't change.
        8. Calculate the amount of suspended solids by using the following formula:

$$\text{Total suspended matter (mg/l)} \times \frac{(\text{weight of ``dirty'' filter}) - (\text{weight of clean filter}) \times 1000}{\text{size of sample filtered (in ml)}}$$

2. Chemical Nature of the Stream
    Use the kits provided to measure the levels of certain chemicals present in the water. Please follow the directions furnished with the kits.
    A. Oxygen concentration (Group II)
        The amount of oxygen in the water is critical for the organisms that live in it. Several factors influence oxygen level.
        1. The organic matter entering a stream from runoff, sewage, and other sources results in lower oxygen concentrations, because the breakdown of organic matter to CO_2 and H_2O is an oxygen-requiring process.
        2. The amount of plant life in the streams can also raise O_2 levels when the plants are photosynthesizing or lower it as they decay.
        3. Finally, the amount of turbulence (tumbling over rocks, etc.) can increase the amount of O_2 able to dissolve in the water. Carefully follow the directions furnished with the kit. Repeat the test until you obtain consistent results. Record results.
    B. Total soluble phosphate (Group III)
        Phosphates may enter streams from several sources.
        1. Fertilizer runoff from lawns and farmland is a major source of phosphates.
        2. Many insecticides are phosphate-containing compounds, which may show up in streams as a result of their use on adjacent lawns and agricultural land.
        3. Other human uses, such as detergents, may also add phosphates to a stream.
        4. Animal waste from farms and outflows from sewage treatment plants also contribute significant amounts of phosphates. Carefully follow the directions furnished with the kit and determine the amounts of soluble phosphates present. Repeat each test until you obtain consistent results. Record results.
    C. Nitrogen compounds (Group IV)
        Nitrogen compounds (ammonia [NH_3], nitrate [$-NO_3$], and nitrite [$-NO_2$]) may enter streams from several sources. Runoff from pasture lands or feedlots can add significant ammonia. Likewise, septic systems that drain into streams can add ammonia. Nitrate and nitrite can also be added from the atmosphere (automobile exhaust contains a variety of nitrogen oxides) or as fertilizer runoff from land. Follow the directions furnished with the nitrogen kits and determine the amount of ammonia, nitrate, and nitrite in the water. Repeat each test until you obtain consistent results. Record results.
3. Biological Nature of the Stream
    A. Bottom organisms (Group V)
        The organisms in a stream are often a good indicator of the quality of the stream. Some species thrive in conditions of low oxygen and high organic matter, whereas others must have water with high oxygen concentrations and low levels of organic material. Collect samples of the bottom, sort through the material, and collect organisms for identification in the lab. Also collect some of the sediment from the bottom to return to the lab, where organisms can be sorted under a microscope. Identify organisms using the keys provided by your instructor.

B. Standard plate count (Group VI)

A standard plate count is a method of assessing the total number of several kinds of bacteria in a water sample. Not all bacteria will grow under the conditions used, but most kinds will and a standard plate count gives an index of bacterial numbers.

1. Use a sterile bottle to collect a sample of water from the stream. Collect away from the bank and avoid collecting sediment from the bottom. The samples should be processed before six hours have elapsed. If a longer transit time is anticipated, the sample should be refrigerated at 10° C or less and in any case should he processed before thirty hours have elapsed. The following procedures should not take place in direct sunlight.

2. To assess the numbers of bacteria present you will need to prepare a series of dilutions of the original sample as follows:

    a. Obtain a sterile water blank containing 99 ml of sterile water.

    b. Use a sterile pipette to transfer 1 ml of the water sample to the 99 ml sterile water blank. Mix the sample with the water in the water blank.

    c. Obtain four sterile petri dishes labeled 1 ml, 0.1 ml, 0.01 ml, and 0.001 ml.

    d. Use a sterile pipette to transfer 1 ml of the original water sample to the petri dish labeled 1 ml. Use the following technique to do so. Lift the lid of the petri dish just enough to allow the pipette to deliver the water to the sterile empty dish.

3. Use the same technique to add water to the other petri dishes as follows.

    a. 0.1 ml from original sample to dish labeled 0.1 ml

    b. 1.0 ml from mixed diluted sample to dish labeled 0.01 ml

    c. 0.1 ml from mixed diluted sample to dish labeled 0.001 ml

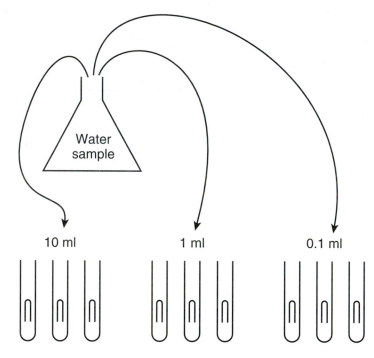

Figure 13.1   Putting the water sample into nine fermentation tubes containing lactose broth.

4. Melt the tryptone glucose extract agar and hold at 44–46° C until used.

5. Lift the lid of each petri dish in turn just high enough to allow you to pour 10–12 ml of the agar into the petri dish. Swirl the contents gently to mix the water sample with the agar.

6. You should also make some petri plates using the sterile water and the medium to see that they were not contaminated.

7. Incubate the petri dishes for 48 hours at 32° C.

8. After 48 hours count the number of bacterial colonies growing on each plate. Since you know the size of the original sample, you should be able to determine the number of bacteria per milliliter of the original water sample. (Ideally, one of your plates should have between 30 and 300 colonies. Use this plate for determining the number of bacteria per ml.)

C. Coliform bacteria (Group VII)

Coliform bacteria are found in the intestines of humans and other animals; therefore, the presence of these kinds of bacteria is an indication of contamination from human or animal waste products. The coliform bacteria themselves are not normally a hazard, but indicate that other pathogenic (disease-causing) bacteria may also be present. If the source of the coliforms is human, then we can assume that some human pathogens will be present.

	Stream 1	Stream 2
Temperature (°C)		
pH		
Total suspended matter (mg/l)		
Oxygen concentration (ppm)		
Total soluble phosphate		
Nitrogen		
Ammonia (mg/l) _____		
Nitrate (mg/l) _____		
Nitrite (mg/l) _____		
Organisms identified		
Standard plate count (number/ml)		
Coliform bacteria (number/100 ml)		

1. Use a sterile bottle to collect a sample of water from the stream away from the bank. Be careful not to collect sediment from the bottom. Stopper the bottle. If the sample cannot be used immediately (within one hour), it should be cooled to 10° C or less until it can be used. In any case, it should be used within six hours.

2. Obtain nine fermentation tubes containing lactose broth. Three should be labeled 10 ml, three should be labeled 1 ml, and three should be labeled 0.1 ml (figure 13.1).

3. Shake the water sample thoroughly.

4. Use a sterile pipette to place 10 ml of the water sample into each of the tubes labeled 10 ml; 1 ml into the 1 ml tubes; and 0.1 ml into the 0.1 ml tubes.

5. Incubate at 35° C for 24 hours.

6. After 24 hours examine the tubes for the presence of gas in the small inverted vial.

7. Examine again at the end of 48 hours.

8. Record all the tubes that show a positive test. A test is positive if gas has collected in the vial *and* the culture is cloudy.

9. Consult a most probable number table to determine the approximate number of coliform bacteria in a 100 ml sample. Your instructor will provide a most probable number table.

VI. Questions

1. List three water-quality differences you think are typical of a polluted stream and an unpolluted stream.

2. Why is it important to determine the number of coliform bacteria in a water source?

3. What kinds of organisms were typical of the polluted stream? Were they different from those found in the unpolluted stream? What does this indicate?

I. Objectives

After observing biological life in a stream, the student will be able to:

1. Understand some of the more fundamental measures of stream quality.
2. Understand some of the factors that can lead to reduced water quality.
3. Investigate the effects of reduced water quality on aquatic life.

II. Introduction

Water is a key component to life throughout the world. Indeed, it is clear that many of the richest and most diverse land-based ecosystems are those that receive a substantial amount of rainfall annually. In aquatic ecosystems, water quality determines the kinds of plants and animals that can inhabit the water. When water quality declines, the number of kinds of organisms is decreased as well. Furthermore, the kinds of organisms that live in poor quality water are considered less desirable than those that live in good quality water. While a chemical analysis of water quality may show high quality at a specific point in time, an analysis of the organisms present tells us something about the kinds of stresses placed on the stream over a year or more. If these stresses were of short duration (pesticide spill, runoff from a heavy rain storm), we would have to chemically sample water quality at the right time to note the effect, whereas sampling organisms is a better indicator of overall, longer term, quality.

Toxicity: Chlorine, acids, heavy metals, pesticides, and other pollutants increase the toxicity of the stream and reduce the numbers of insects and fish in the water. A toxic problem is typically the only reason a water system will be totally lacking insects.

Physical: Physical conditions such as heat and sedimentation can reduce the quality of the stream for insects and fish. Increased temperatures reduce the amount of oxygen in the water, making it difficult for some insects and fish to live. Increased amounts of sediments in the water can lead to less visibility, and increased silting in the streambed. This reduces feeding opportunities for organisms that use vision to locate food. Silting can also cover and destroy valuable rocky and pebbled streambed habitats that supply hiding places for many kinds of insects and provide sites for fish to lay eggs.

Organic: Organic pollution such as livestock wastes, agricultural fertilizers, and human wastes can have drastic effects on stream quality. Organic pollution usually decreases the amount of oxygen available in the water, thereby changing the stream so that it becomes dominated by life-forms such as trash fish, worms, and bacteria, which can survive low-oxygen concentration.

III. Relevance: Biological Indicators

Water pollution exists in many forms. Some types, such as excessive sediment from erosion, or the dumping of old tires, is easily seen. Other forms of pollution, such as thermal (heat) or chemical (from pesticides or other compounds), are not easily perceived or detected, at least in the early stages of pollution. Biological organisms, such as insects or fish, are often the earliest indicators that the water has been impaired in some way, although the effects may not be obvious to the human eye. The initial condition of a stream can be quickly and cheaply determined simply by observing what types of organisms live in the water. Some species, such as mayflies and dragonflies, are extremely sensitive to polluted water, so that their absence may indicate impaired water quality. Others, such as some species of flies, are more tolerant of pollution, and their numbers may actually increase in a polluted water system. The absence of any life-forms, even if the water seems unpolluted, may indicate severe forms of contamination. If the biological indicators show potential pollution, more sophisticated chemical tests can be run to determine the exact cause of the pollution.

IV. Activity

In this exercise, you will take a field trip to a local stream and determine the quality of the water based on its biological inhabitants. Because organisms have a particular range of tolerance for various chemicals in the water that they can live in, you will be able to judge, simply by the presence or absence of particular species, the quality of the stream. Immature insects, called **benthic macroinvertebrates**, are good indicators of water quality. These insects, which burrow into the soil or live under rocks in the stream, are easily identified, usually numerous (unless the water is particularly polluted!), and are generally harmless to handle (although some species do have pincers, so be careful when handling them). They are, however, relatively fragile creatures, so you should be very careful when collecting and identifying them, so that you don't harm or kill them. Of course, when you are done with them, return them to the water, since many other species (such as fish and birds) depend on them as a food source. Once you have observed, counted, and identified the various macroinvertebrates, discuss among yourselves which species seemed most numerous, and the possible reasons for the different abundances of the species. If someone in your class is particularly interested, or proficient, in identifying the species of the organisms (compared to just classifying them by color, size, etc.), there are a number of very good, easy-to-use field manuals for identifying benthic macroinvertebrates.

V. Procedure

Materials
Per group:

> Needle-nose forceps
>
> White collecting pan
>
> Several jars of different sizes
>
> An old window screen with no holes, approximately 1 meter by 1 meter
>
> Old tennis shoes for wading in streams
>
> Magnifying glass

Method
Students will work in groups of two.

1. Position the screen vertically at right angles to the flow of water so that the water flows through the screen. If the stream has a firm, sandy bottom, place the screen so that it is downstream from an area of the stream bottom that is covered in part with sand, and in part with leaves, sticks, or other debris. Be certain that the bottom edge of the screen is flush with the bottom of the stream, allowing no insects to escape. Also, do not allow any water to flow over the top of the screen, as this could also allow insects to escape.

2. Measure a distance 1 meter upstream from the screen. In this area, pick up the rocks, brush them with your hands and let the loose organisms, debris, and other matter float down into the screen. If the bottom is sandy or muddy, agitate this material lightly by kicking the streambed in a diagonal motion toward the screen for 30 seconds. After kicking the streambed, wait 2 minutes while the water flow carries the debris into the screen.

3. Once you have waited the 2 minutes to allow material to collect in your screen, pull the screen out of the stream so as to keep the material collected on the screen. Put the screen on a light-colored, flat surface and pull all the material from the screen with your forceps and place it in the white collecting pan containing sufficient water to allow the animals to survive. Look very closely at the material in the collecting pan and record any objects that are moving animals. Most of the organisms will be insects and other arthropods, although some worms and snails may be collected. We will concentrate on the insects, which will be the organisms with three body segments and six legs.

4. Once you are certain that you have removed all the insects from the screen, count the total number you have collected. Separate the insects into groups that look generally similar in terms of body style, number of legs, wings, or tail shape. Place the separated groups into collecting jars that are also partially filled with stream water.

5. To evaluate the quality of the stream, we will tabulate the numbers of insects you found by recording in Data Sheet 14.1:
   a. The total number of insects captured.
   b. The total number of individual insects that have bodies that are white, red, or gray. Do not consider the color of the heads, only the body.
   c. The total number of individual insects that have bodies that are black, green, brown, or tan. Ignore the color of the head.
   d. The total number of different kinds (species) of insects that you can identify.
   e. The total number of the most common species of insect.

6. Record the presence or absence of fish seen as you walk along 30 meters of the stream.

Stream Quality Categories

Dead:
A. Less than one organism found per square meter sampled.
B. No fish observed.

> *Note:* Clear water, with nothing living in the water, suggests the stream has a toxicity problem.

Poor:
A. More than one insect found per square meter.
B. 90% of organisms found are gray, white, or red in color.
C. One to two types of organisms are distinguishable.
D. No fish observed.

> *Note:* The low level of organisms and the lack of variety in the kinds of organisms suggests that only the most resistant species can live in the water.

Fair:

A. More than one insect found per square meter.
B. 11–30% of organisms found are black, green, brown, or tan in color.
C. Three to six types of organisms are distinguishable.
D. There is not an organism that makes up more than 90% of the sample.
E. Fish are observed.

Note: Some of the species found require good quality water. Somewhat reduced diversity indicates either past problems from which the stream is recovering, or low levels of contamination.

Good:

A. More than one insect found per square meter.
B. At least 30% of organisms found are black, green, brown, or tan in color.
C. Six or more types of organisms are distinguishable.
D. There is not an organism that makes up more than 50% of the sample.
E. Fish are observed.

Note: The species diversity and types of species present indicate constant high-quality water, and the lack of contamination.

Data Sheet 14.1:
Stream Quality Assessment

Name _____

Section _____

Date _____

Characteristic	Number	Percent of total
Total number of insects collected		XXXX
Number of different kinds (species) of insects identified		XXXX
Number of kinds that are white, red, or gray	white _____ red _____ gray _____ Total 1 _____	$\left(\dfrac{\text{Total 1}}{\text{Total}}\right)$
Number of kinds that are black, green, brown, or tan	black _____ green _____ brown _____ tan _____ Total 2 _____	$\left(\dfrac{\text{Total 2}}{\text{Total}}\right)$
Total number of the most common species of insects collected	(Total 1 & Total 2) Total = _____	XXXX
Fish observed?	Yes No	XXXXXXXX

VI. Questions

1. What do the organisms collected tell you about the quality of the stream?

2. Would you have agreed with the results in question (1) by simply looking at the stream? Why or why not?

3. List three activities that take place on land that could affect the quality of the water in the stream.

4. What are some common pollution control techniques used to improve stream quality?

Air Pollution

I. Objectives

After completing the work associated with this exercise, the student will:

1. Understand the significance of particulate pollutants.
2. Become familiar with an air sampling technique for total suspended particulates (TSP).
3. Appreciate the role of weather in air pollution problems.
4. Determine whether his or her environment is free of air pollutants.

II. Introduction

Like water pollution, air pollution is becoming an increasing problem in our society. The detrimental effects of air pollution—poor visibility, eye and throat irritation, and damage to vegetation and property—are obvious to anyone who has ever visited or lived in or near a major urban area. Several U.S. cities (notably Los Angeles, Denver, and Houston) are already at the point of having to take decisive action to improve air quality, or risk subjecting their citizens to significant health risks.

Air pollution is defined as the release of harmful amounts of natural or synthetic materials into the atmosphere as a direct or indirect result of human activity. The severity of air pollution in a given area depends on several factors, including climate, topography, population density, and the number and type of industrial activities. Air pollution may be broken down into four categories. The first, **ambient,** or outdoor air pollution, is perhaps the most familiar and includes both natural (biogenic) and human-made (anthropogenic) sources. However, **indoor** air pollution is also of concern because, on average, most people spend up to 90 percent of a typical day indoors. **Occupational** air pollution is restricted to the workplace, but can involve potentially heavy exposures, often to more than one pollutant. Finally, **personal** air pollution includes voluntary exposures to consumer products, as well as those incurred through lifestyle choices (such as smoking).

Increases in ambient air pollution are due primarily to four factors: (1) population growth; (2) technological changes that create new products; (3) social changes (such as urbanization); and (4) a rising standard of living. In general, as energy use per person increases, so too does the potential for increased air pollution. Most ambient air pollutants are added directly to the troposphere, the lower layer of the atmosphere, as **primary pollutants**. In the troposphere, and in the presence of sunlight, they often mix and react chemically with each other or with natural components of the atmosphere, forming **secondary pollutants**. Eventually most of these pollutants are washed out of the atmosphere, and returned to both land and water by precipitation. Some contaminants, such as large particulates, settle out under the influence of gravity as fallout. Insoluble and nonreactive chemicals, on the other hand, may diffuse upward into the stratosphere, adversely impacting the ozone layer or contributing to global climate change.

In the United States, ambient air pollution is regulated at the federal level under the Clean Air Act (CAA), which establishes standards for the following classes of air pollutants: carbon oxides, sulfur oxides, nitrogen oxides (NO_x), photochemical oxidants (ozone), and particulate matter (PM). **Particulate matter** represents approximately 5 percent of all air pollutants by weight, and includes a variety of components such as sulfate salts, sulfuric acid droplets, metallic salts, dust, and liquid sprays and mists. Particles range in size from 0.005 micrometers (or microns) to greater than 100 micrometers. (*Note:* 1 micrometer = 1 μm = .001 mm). Biogenic sources of particulates include wind erosion, various pollens, volcanic eruptions, and forest fires caused by lightning. Particles derived from these sources are usually large and settle out rapidly. Anthropogenic sources of particulates are generally associated with one of three activities. **Attrition** adds particulates to the air by sanding, grinding, drilling, and spraying (as in a manufacturing process). **Vaporization** is the phase change from a liquid to a gas and is a major source of nuisance odors in the air. **Combustion**, or burning, is a conversion process, which results in the chemical combination of specific substances (fuels) with oxygen. Because fuel burning activities are so widespread, combustion accounts for the vast majority of particulate generated. Major particulate sources include internal combustion engines, coal-fired power plants, boilers and industrial furnaces, and solid/hazardous waste incinerators.

III. Relevance: Regulatory Standards for Particulate Matter

The environmental and health effects of particulates vary according to their size. For example, very small aerosols act as nuclei on which water vapor condenses, which increases the frequency and duration of fog and ground mists. Particle size also influences how deeply a particle may penetrate into the human respiratory system. Larger particles (>7–10 microns) are filtered out by nasal hairs (cilia) and cleared by blowing the nose, or are washed into the gastrointestinal tract by mucous membrane action. Smaller particles penetrate much deeper into the respiratory system. Most are cleared from the lungs, but particulates that reach the alveoli—small saclike structures deep within the lung across which gas diffusion takes place—are often trapped and retained there indefinitely.

During the 1970s and 1980s, air-quality standards for particulate matter were based on total suspended particulate (TSP) measurements obtained using a high-volume air sampler. In the case of TSP sampling, the size range of particles trapped on the filter is typically less than 25–50 μm, but greater than 10 μm. However, subsequent research into the health effects of particulate matter concluded that the particle size cutoff of the high-volume sampler was significantly greater than that of the human respiratory system. In other words, the TSP method measured particulates that were removed by natural clearing mechanisms in the upper respiratory system, and therefore never reached the lungs! This finding resulted in a new regulatory standard for particulates less than 10 μm in diameter, known as the PM_{10} standard. The 24-hour standard for PM_{10} is 150 μg/m^3.

IV. Activity

In this exercise, you will use a high-volume ("hi-vol") air sampler to measure TSP. The 24-hour air-quality standard established for TSP is 260 μg/m^3. Because of the expense involved, PM_{10} samplers will not be used, but the operating principle for the two types of air samplers is the same. Both devices are designed to draw air at a constant rate through a glass fiber filter, which traps the particulate matter (figure 15.1). The weight of particulate matter collected can be determined by weighing the filter before and after sampling. The filter should be handled only with forceps, since dirt and oils from your skin will increase its weight. To eliminate any weight increase due to humidity, the filter should be desiccated (dried) for at least 24 hours before and after the sampling run. Since the hi-vol has a constant flow rate, the volume of air sampled can be determined if the sampling time is known.

Most high-volume samplers have a flow rate of 1 cubic meter per minute (1 m^3/min). Be sure that the sampler you are using is correctly calibrated and remember to check the manufacturer's specifications. By using the raw data obtained, and a few simple equations, you can determine the concentration of airborne particulates (C_p) present on the sampling days. The C_p is expressed in μg/m^3 (micrograms per cubic meter).

In its entirety, this lab requires four days to complete. The lab consists of two independent parts: using a hi-vol air sampler to measure particulate concentrations over a 24-hour period, and obtaining weather observations to see how the sampling results correlate with any weather changes. However, if time is an issue, or if your school does not have the necessary equipment, the lab can be modified quite easily to fit your school's resources. If the entire lab is done, each student is expected to come in on one other day outside of scheduled class time. The time schedule is explained further in the Procedure section.

A typical air sampler device

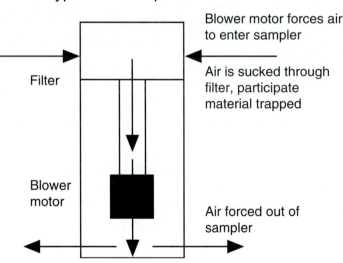

Figure 15.1   A typical air sampler device

V. Procedure

Materials

Part 1: Hi-vol air sampler:

forceps
8″ × 10″ glass microfiber filter
envelope
desiccator
Mettler balance
high-volume air sampler
calculator
data sheets

Part 2: Air measurements:

max/min outdoor thermometer
handheld anemometer
compass or wind vane
hygrometer or sling psychrometer
barometer
rain gauge
radio

Methods

For both Part 1 and Part 2: perform all of the following steps.

Part 1 only: Perform all the following steps EXCEPT step 9.

Part 2 only: Perform ONLY step 9.

The week *before* this lab, the class will divide into three groups:

1. Group 1 will place the filter in the desiccator 24 hours *before* the lab.
2. The entire class will remove the filter from the desiccator, weigh it, and place it in the hi-vol sampler *during* lab.
3. Group 2 will remove the filter from the hi-vol sampler 24 hours *after* lab and place it in the desiccator.
4. Group 3 will remove the filter from the desiccator and weigh it 24 hours *after* it was placed in the desiccator.
   (*Note:* If there are multiple lab sections, it helps to have two or three hi-vol samplers.)

24 hours BEFORE lab—Group 1:

5. Using forceps, carefully remove a filter from its storage box and write your lab section in pencil below the identification number. Using the forceps, fold the filter into quarters and place it in an envelope. The envelope may now be handled without contaminating the filter.
6. Place the envelope containing the filter into the desiccator for 24 hours. (*Note:* Be sure that the desiccant is adequately "charged." Most desiccants will show a color change, which indicates that the desiccant needs to be regenerated or replaced.)

In lab—Entire class:

7. Remove the envelope from the desiccator and, using forceps, carefully remove the filter from the envelope and weigh it on the Mettler balance. Record the initial weight (W_i) on Data Sheet 15.1.
8. Using forceps, remove the filter from the balance and place it, unfolded, numbered-side down, in the frame of the hi-vol. Clamp it down, close the hi-vol canopy and turn it on. Record the start time on Data Sheet 15.1.
9. After starting the hi-vol, make some observations on the weather conditions. If you have access to the instruments listed above, your instructor will advise you on how to use them. In the event that such instruments are unavailable, simply make notes regarding cloud cover, wind direction, wind speed, precipitation, and so on. You can use the Beaufort Scale to estimate wind speeds (see Appendix C). Supplement your notes with meteorological data from a local newspaper or weather forecast on radio or television. You may also use a weather radio or telephone for up-to-the-minute information on area conditions. Record your observations on Data Sheet 15.2.

Recording weather observations:

1a. Cloud cover: Estimate the percent of sky that is covered with clouds. Record the following according to your estimate:

<10% cloud cover = Clear (no clouds)
20%–50% cloud cover = Scattered clouds
50%–90% cloud cover = Broken clouds
>90% cloud cover = Overcast

1b. If there are clouds, note the approximate height (low, medium, or high), type, and direction they are moving (with a compass).
2. Record the weather conditions, using the following codes:

NP = no precipitation
D = drizzle
MR = moderate rain
T = thunderstorm
F = fog

3. Record the temperature.
4. Record the relative humidity (as a percentage), by following the directions of your psychrometer or hygrometer.
5. Record the barometric pressure (in millibars) of the day. Compare the pressure to a reading taken earlier in the day. If you have noted a rise or fall in air pressure during the previous three hours, mark a + or − accordingly (your instructor or a previous class will have taken a reading earlier in the day).
6. Record the wind direction by using the Beaufort Scale (Appendix C) by entering the code that best matches the conditions of the day.
7. Record precipitation, if any, to the nearest 0.01 inch by using a rain gauge. If no precipitation has occurred, enter "NP." If precipitation is less than 0.01 inch, enter "T" for trace.
8. Enter any special phenomena you observed such as thunder and lightning, smoke, or haze.

24 hours AFTER lab—Group 2:

10. Turn off the hi-vol, record the time and, using forceps, remove the filter. Place the filter back into the envelope, folding the sides so no particulates are lost into the envelope, and desiccate for 24 hours.

24 hours later—Group 3:

11. Remove the envelope from the desiccator and, using forceps, remove the filter from the envelope, and weigh it on the Mettler balance. Record the final weight (W_f) in the appropriate row and column on the table on Data Sheet 15.1.
12. Complete the calculations to determine the concentration of particulates (C_p). Record all data on Data Sheet 15.1.

Data Sheet 15.1

Name _____

Section _____

Date _____

Group number: _____

A. Complete the following table:

Parameter	Filter
Initial weight (W_i in g)	
Final weight (W_f in g)	
Day and time the filter is put on the hi-vol sampler	
Day and time the filter is taken off the hi-vol sampler	
Total time sampled (in min.)	
Hi-vol flow rate (check manufacturer's specifications)	
Volume of air sampled (m^3/min)	
Mass of particulates ($M_p(\mu g)$)	

B. Complete the following calculations for each of the four filters:
   1. To determine the mass of particulate matter collected (M_p) in micrograms (μg), use the following formula:

$$M_p(\mu g) = (W_f - W_i) \times (10^6 \ \mu g/g)$$

   2. To determine the volume of air sampled, use the following formula:

$$\text{Volume (m}^3) = \text{Total time sampled (min.)} \times \text{machine flow rate (m}^3/\text{min})$$

   3. To determine the concentration of particulates (C_p), divide the mass of particulates ($M_p(\mu g)$) by the volume of air sampled (m^3):

$$C_p = M_p(\mu g)/m^3$$

Data Sheet 15.2

Name _____

Section _____

Date _____

Group number: _____

A. Record your surface observations using the chart below:

Variable	Day
Time/date of Observation	
Cloud cover	
Cloud height	
Weather conditions	
Temperature-min (°F)	
Temperature-max (°F)	
Rel. Humidity (%)	
Barometric pressure (mbar)	
Wind direction/speed (mph)	
Rainfall (in.)	

Notes:

_____

_____

_____

_____

_____

_____

_____

_____

_____

_____

_____

_____

_____

_____

_____

_____

_____

VI. Questions

1. Describe the appearance of your filter following the sampling period. Did you expect the filter to look like this?

2. Did you observe any weather changes during the experiment? If so, how might this have affected your results?

3. Did any of your airborne particulate matter concentrations exceed 260 $\mu g/m^3$ in a 24-hour period? If so, what should be done about it?

4. How would your filter look if it were placed alongside a road? In the country? Next to a factory? Why is the location of the filter important?

5. Do you consider the air pollution in your area to be dangerous? If so, what are the contributors to air pollution in your area? If air pollution is not a problem now, do you think it will be in the future if nothing is done to protect air quality?

I. Objectives

After completing an analysis of soil in this exercise, the student will be able to:

1. Understand characteristics of soil, its formation, and importance.
2. Classify a given soil sample based on its texture, water-holding capacity, color, and amount of organic matter.
3. Conduct chemical analysis of soil horizons using commercially available soil testing kits.
4. Discuss critically the suitability of a given soil for agricultural cultivation or landfill location, etc.

II. Introduction

Soil is defined as the unconsolidated mineral material located on the surface of the earth. Soil is formed by the physical (e.g., freezing, burrowing of animals, or the grinding action of water) and chemical (e.g., dissolving of rocks by acid, water, etc.) weathering of parent material or bedrock.

After soil is formed, it may be transported from its site of formation by water (streams, rivers, lakes, and storm runoff), wind, gravitational forces (mudslides), and glacial ice. As a result of soil formation and transportation, soil horizons develop. **Soil horizons** are the layers of soil in a given area that differ by physical structure as well as biological and chemical content. If one were to cut a section vertically downward through the soil, the various horizons would be visible. This vertical section is called a **soil profile**. Well-developed soil has its own distinctive profile, with characteristics related to the nature of the horizons, the degree of weatherization, and the percentage of clay and organic matter within each horizon.

While there are many physical properties of soil that can be measured, four very important characteristics are soil texture, water-holding capacity, color, and organic content. **Soil texture** is determined by the proportion of soil particles of varying sizes. The primary sizes in decreasing order of coarseness are: sands, silts, and clays. The relative proportion of these particles determines a soil's texture. A soil's texture affects water absorption and retention, aeration, and the fertility of the soil. Sandy soils drain well, are well aerated, and easy to till, but they also dry rapidly and lose vital nutrients as water drains through the soil. They are also very loose soils and are poor for building on. Clay soils have very small, tightly fitted particles. There is very little pore space and the soil is difficult to wet, drain, and till. They are, however, more stable. Silty soils drain well, are easy to till, and usually contain abundant nutrients that make them good for growing crops.

Water-holding capacity is important for plant growth and soil stability. The strength with which soil holds water is important to plant growth. Water in clay soils is tightly bonded and is not the most suitable medium for plant growth. Sand is not a suitable medium because its large size has a smaller surface area to volume ratio and does not retain much of the water absorbed by the soil. An even mix of clay, silt, and sand, known as loam, is the most suitable soil type for plant growth.

Soil color can indicate a soil's stage of development or mineral origin. White colors indicate the presence of salts or carbonate deposits in the soil. Mottled colors indicate that a soil has periods of poor aeration. Blue, gray, or green tinged subsoils indicate long water-logged periods resulting in poor aeration. Dark colors usually denote soils containing high amounts of organic matter. (Table 16.1)

Table 16.1 Characteristics of soil, based on soil color

	Dark (Dark gray, brown, black)	*Mod. Dark* (Dark brown to yellow-brown)	*Light* (Pale brown to yellow)
Amount of organic matter	Excellent	Good	Low
Erosion factor	Low	Medium	High
Available nitrogen	Excellent	Good	Low
Fertility	Excellent	Good	Low

The soil's **organic content** increases its water-retention forces and nutrient levels. Organic matter also lowers the rate of soil erosion, and provides for better aeration and water movement in the soil. Organic matter can occur naturally in a soil as detritus (decomposed plant and animal material), or it can be added to a soil in the form of composted material.

III. Relevance: Soil Erosion

Humans often make poor land-planning decisions, which have long-lasting, negative effects. In rural settings, farmers remove the biomass, thereby degrading the soil, reducing its nutrients, and increasing erosion. Heavy farm equipment compacts soil, reducing its air and water-holding capacity. Inorganic fertilizers and pesticides can build up in soils, and disturb their natural balance. In urban settings, new construction removes most vegetation, which results in increased erosion. Homes, factories, roads, and landfills reduce the availability of land for food production.

Although erosion is a natural process, actions such as overgrazing and deforestation accelerate this problem to great proportions. Erosion is a problem in countries like the United States, because many people do not concern themselves with soil conservation practices. It was the clearing of native grasslands for the growing of wheat that led to the great Dust Bowl of the 1930s. In Third World countries, poverty and overpopulation drive people to use the land beyond its capacity for recovery in order to eat for one more day. Worldwide, an area the size of Brazil has become desertified in the last 50 years because of soil loss. Every inch of topsoil that is lost results in a 6 percent loss in agricultural productivity. Thus, each year we have to feed millions more people with less topsoil. The goals of soil management are to prevent erosion and nutrient depletion. Farmers can prevent most erosion by using minimum tillage, contour farming, strip cropping, and terracing. Nutrient depletion can be prevented by using crop rotation, organic farming, and polyculture practices.

IV. Activity

In this activity we will go to the field and measure the slope of the area, take soil probe samples, and sketch the soil profile. We will observe characteristics such as soil color, texture, depth, and water-holding capacity. We will also test the pH, nitrogen, potassium, and potash levels of the soil.

V. Procedure

Materials (per group)

    1 meterstick

    1 centimeter ruler

    1 leveling instrument

    1 soil probe

    1 Sudbury soil test kit

    soil sample collecting containers (small plastic bags with twist ties)

    labels to put inside each baggie

    data sheets

    1 funnel

Method

1. Divide the students into groups of 2 to 3.
2. Go to an area of land where you would like to determine its most valuable use. Answer the questions on Data Sheet 16.1, section A.

Each group will do the following:

3. To measure the average slope of the area:
    a. Select a site that represents the average slope of the land.
    b. Hold the meterstick parallel to the ground, with one end touching the ground and the other end pointing downslope. Hold the stick outright until it is approximately level. Place a leveling instrument on the stick and raise or lower the meterstick until the bubble is centered.
    c. Using the centimeter ruler, measure the number of centimeters between the free end of the stick and the ground.
    d. The number of centimeters on the centimeter ruler equals the slope of the land in percent. Record the percent of slope on Data Sheet 16.1.

4. With the soil probe, gather a soil sample. Immediately measure the depth of each distinct layer. On Data Sheet 16.2, draw and label the soil layers and depth. Use Figure 16.1 for help in identifying the layers.
5. Divide the sample by its different horizons and put each in a separate container. Bring these samples back to the lab for further analysis.

6. For each layer, complete the soil characteristics chart on Data Sheet 16.1. Use the following information to identify the characteristics of each of the layers:

    Texture: Moisten the soil and squeeze it between your thumb and forefinger.

If it feels:	It is:
gritty	sand
smooth, slick, not sticky	silt
smooth, plastic, sticky	clay

    Water-holding capacity: Fill the funnel with a soil sample, and pour water into the funnel. Watch the rate at which the water passes through the soil.

If it passes through:	It is:
easily	sand
more slowly	silt
very slowly	clay

    Color: Use the color chart in the Introduction section, and record whether the sample is dark, moderately dark, or light.

    Amount of organic matter: Look at each layer. Approximate the percentage of organic matter. Record whether the sample is high, medium, or low in organic matter.

7. Each group will determine the chemical composition (nitrogen, phosphorus, potash, and pH) of the first three layers of soil. Follow the directions that come with your soil testing kits. Record data on the chemical composition chart of Data Sheet 16.1.

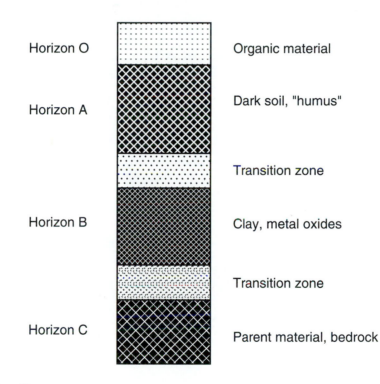

Figure 16.1   A typical soil profile

Data Sheet 16.1

Name _____

Section _____

Date _____

Group number: _____

A. Complete the following description of your soil study area.
   1. What is the topography of your study area (hilly, flat, rocky, etc.)?

   2. How is this land and surrounding land currently being used?

   3. What type of biomass is the soil currently supporting?

   4. Do there seem to be any soil-erosion problems in the area? If so, what kind, and what appears to be the cause?

   5. What is the slope of your study area? _____ percent

B. Soil Characteristics Chart

Layer	Texture	Water-holding capacity	Color	Organic matter
Top 1				
2				
3				

C. Chemical Composition Chart

Layer	Nitrogen	Phosphorus	Potash	pH
1				
2				
3				

Data Sheet 16.2

Name _____

Section _____

Date _____

Depth   Horizons

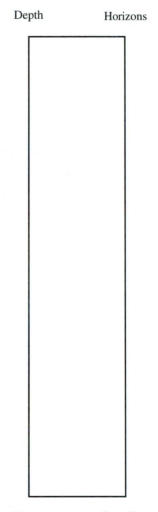

Draw your own soil profile

VI. Questions

1. How does the slope of your soil survey area affect use of the land?

2. Is the soil texture in the area you surveyed best suited for cultivation or for a landfill site? (Hint: Is the texture mostly sand, silt, or clay?)

3. What other uses would the land be suitable for? Take into account soil type, environmental condition, and location.

4. Does the nitrogen level in your top soil accurately depict what the soil color chart in the Introduction section predicted?

5. From the chemical composition of your sample, what type of fertilizer would you recommend, if any, for your survey area if it is to be used for cultivation?

6. What improvements to the survey area would you suggest to increase its use value and decrease the amount of environmental degradation?

Field Trip Suggestions

1. Visit a local industry's water pollution-control facilities. Determine the cost of operating the facility. What are the pollution problems the industry is trying to control? What techniques are used?
2. Canoe down a local stream or river through an industrial area, and record examples of pollution. Consult with your local pollution-control agency about your findings. Do they agree or disagree that the examples you cite are important?
3. Fly over or drive through an industrial area, and record examples of pollution. Consult with your local pollution-control agency about your impressions. Do they agree or disagree that the examples you cite are important?
4. Visit your local water treatment plant. Make a list of the treatment methods employed, and what purpose each method accomplishes. Ask what the major water-quality problems are in your area.
5. Visit a farm and record five examples of good soil-conservation practices and five examples of poor soil-conservation practices.

Alternative Learning Activities

1. Dig a hole or observe a road cut and identify the different layers of a soil profile.
2. Ask a spokesperson from your local water planning board to talk to the class about where water is used and how water use will affect the economy of the area.
3. Invite a spokesperson from an environmental group to talk to the class about the major pollution problems in your area.
4. Collect newspaper articles that relate to pollution. Note the sources used in writing the article. If the article is by a local writer, phone him or her and try to determine what he or she knows about the subject, or ask the writer to come to class to discuss how information was gathered for the article. Choose five articles and comment on the validity of the impressions they give.
5. Have a person from a local pollution-control agency visit the class and discuss its function.
6. Do an inventory of hazardous chemicals in your home. Determine the proper disposal method for each item found.
7. Draw a diagram tracing the flow of wastewater from your toilet to its discharge into a local body of water.
8. Sit quietly in a room, park, or natural area. Identify the sounds you hear. Rank them as to which are the loudest and which are the most annoying.
9. Read *The Tragedy of the Commons* by Garrett Hardin and apply his concept to a local situation. Write a paragraph showing how the local situation is related to Hardin's essay.
10. Capture rain or snow and measure its pH.
11. Maintain several bottles of pond water in sunlight. Add different amounts of fertilizer to each bottle and compare how they differ in appearance over time.

PART 4
Energy Use

Much of human society, especially in the developed countries, depends on energy for its survival. The major sources of energy come from nonrenewable resources: coal, petroleum, and natural gas. Unfortunately, these sources are finite, and will eventually become too rare and expensive to use economically. These resources don't even have to actually run out before alternative sources of energy will have to be found. There will come a time when it costs too much, environmentally and economically, to extract nonrenewable resources from the earth. Before we reach that point, society must determine what alternative forms of energy will be acceptable. Technology, again, will provide many answers, but it will not be enough. The easiest, and fastest, method to implement is to increase the efficiency with which we currently use nonrenewable resources. For example, cars can be designed to get improved gas mileage, houses can be better insulated, and people can obtain their food from local sources and eat lower on the food chain.

Part 4 addresses energy issues. Exercise 17 describes individual energy consumption in economic terms. Exercise 18 examines the use of solar energy to heat water, while exercise 19 discusses various forms of insulation and their relative efficiencies. Finally, exercise 20 examines the personal energy consumption of the students, in order to increase their awareness of their everyday impacts on the environment.

Economics of Energy Consumption

I. Objectives

After monitoring his or her energy consumption, the student will:

1. Become aware of energy consumption information available to consumers.
2. Be able to evaluate the long-term costs of energy-consuming products.
3. Be able to evaluate both energy and economic losses in his or her own home.
4. Assess his/her own contributions to environmental problems and solutions.

II. Introduction

Energy is an important part of the lives of many people; some might say it is indispensable. Energy comes in many different forms, including oil, coal, natural gas, electricity, and hot water. Each one of these forms of energy has different costs associated with it, depending on its supply, ease of distribution, and demand. Many of our energy resources are nonrenewable, thus inefficient use of energy not only decreases supplies, but can also needlessly increase costs to the consumer. Worldwide energy consumption differs greatly among different nations. The United States leads the way. The United States makes up only 6 percent of the world's population, yet Americans use 25 percent of the world's resources. On a per person basis, the average American uses twice as much energy as someone from Japan or western Europe, and 16 times more than a person in a developing nation. Clearly, it is important that we use those resources judiciously and responsibly, both for ourselves and the world. There are many ways you can help solve environmental problems. Two of these ways are rethinking your role and reducing what you use. Think about the ways you, the individual consumer, use energy. Do you let the water run when you brush your teeth? Do you drive someplace you could easily walk to? Do you leave lights on in a room when you are not there? These are but a few questions you should ask yourself. If you find you use energy inefficiently, you can correct your habits *without a major change in your lifestyle*. For example, in areas where curbside recycling has been implemented, some citizens complained at first because they thought it would be too much trouble saving and sorting recyclable materials. But before long, it simply became part of their weekly routine, and the impact on their lives was negligible. Not only will changing the ways you use energy help the environment, you will see savings on your electric bill, water bill, how much you spend on gasoline, etc. Small changes can definitely reap major benefits.

III. Relevance: The Oil Embargo and the Gulf War

The United States has been dependent on oil, especially oil from other countries, for a long time. Prior to 1973, the United States imported over 50 percent of the oil it used. Then in 1973, oil-producing nations imposed an embargo, a restriction in sales, of oil to the United States. The cutback was only 5 percent, but the impact on the United States was tremendous. Prices of gasoline and other petroleum products increased sharply. Gas stations ran out of gas to sell for days at a time because shipments were less frequent. When gas did arrive, people stood in long lines to fill up their gas-guzzling cars before the stations ran out again. In response to this crisis, actions were taken at all levels. Individual citizens started driving less, and walking, bicycling, and car pooling more. The automobile manufacturers designed smaller, less expensive, and more fuel-efficient cars. Oil companies started refining oil from U.S. oil deposits. The government reduced the speed limit to 55 mph. The result was that for 10 to 15 years, overall U.S. demand for oil dropped, and importation of foreign oil was only about 30 percent. By the mid-1980s, however, the United States went back to its old ways. People started driving more, cars became bigger and less fuel efficient, the speed limit was raised, and oil imports were again over 50 percent. In 1990, Saddam Hussein, president of Iraq, invaded Kuwait in an attempt to restrict oil production and raise oil prices. Ultimately, the United States rallied the rest of the world to evict Hussein from Kuwait. The result was a brief, but very destructive and costly, war. Many people said the war was about oil and the freedom to be able to drive a big car. Some think the war was about power, and letting one man gain too much power. It just so happens that because of the world's dependence on oil, oil *is* power, and many were willing to go to war over it. Since the Gulf War, nothing has changed. Oil imports are still high, and cars are still big. How far will we be willing to go next time?

IV. Activity

In this exercise, you will monitor your own energy consumption for one week, and calculate both energy and economic costs of that consumption. In addition to monitoring your own energy usage, you will also do some comparison shopping and calculations to see if the appliances you use, the car you drive, the light bulbs you use, etc., are the most efficient on the market. This will allow you to see if, in addition to changing the ways you use energy, changing the devices you use will help as well.

V. Procedure

Note: if there are time constraints that do not allow the student to do all of the following activities, the class can divide into groups. Each group will then be responsible for gathering data on: (1) lights, (2) major appliances, (3) heating water, and (4) transportation. In class, the groups can then pool their data, and fill in the data sheets. An alternative idea can be for the students to gather information for a shorter time period (e.g., 2 hours), and then extrapolate this information to provide data on an entire 24-hour period. For example, if lights are left on for 1 out of 2 hours, then the data could be extrapolated to mean that, during periods of darkness, the light is on 50 percent of the time.

Fill in the following information on Data Sheets 17.1 and 17.2.

Lights: Check the wattage of the light bulbs in your dormitory, apartment, or home. A 100-watt incandescent bulb gives off as much light as a 40-watt fluorescent bulb. Monitor how many minutes you use the various lights. Calculate the kilowatt-hours using the following formula:

$$\frac{(\text{watts})(\text{total minutes used}/60)}{1000}$$

Now assume that you will need to replace each incandescent bulb once a year, but each fluorescent bulb only needs replacing every 2 years. Price the two types of bulbs and calculate how much you would spend over a 10-year period on each type of bulb.

Major appliances: Record the wattage of your refrigerator, stove, microwave oven, television, wash machine, clothes dryer, CD player, and computer. Record how many minutes you use each appliance and calculate the kilowatt-hours used, using the same equation you used for the light bulbs. Note that a refrigerator uses electricity "on its own," so record how many times it comes on and the length of time it stays on. Next, go to an appliance store and find comparable models of stove and refrigerator. Record both the price and energy usage information. The energy usage information is a yellow sticker on the appliance that tells you how much it costs, on average, to use the appliance each year, based on its energy efficiency, cost of manufacture, etc. Compare that information with one other model of each. Also make a comparison between gas stoves and electric stoves. Calculate the cost of operating each type for 10 years.

Heating water: Heat is measured in British thermal units (BTUs). It takes one BTU of energy to raise the temperature of 1 pound of water just one degree Fahrenheit. Because it takes so much energy to heat water, water heaters are very expensive to operate. Turn on your faucet so that it drips about 1 drip/second. Collect this water for 15 minutes and weigh it. Multiply this number by 35,040 to find the amount of water lost in one year by a dripping faucet. Multiply the amount lost by 80, the average number of degrees Fahrenheit water temperature is raised by a heater. This final number is the number of BTUs used per year just by not turning off the faucet all the way.

Transportation: Record the number of miles you travel by car, foot, bicycle, and public transportation for one week. Calculate your weekly cost of each mode of transportation, such as gas, bus fare, air for bicycle tires, etc. Next, go to a car dealership and find your make and model of car, and then its opposite. If you drive a big car or truck, find a compact car, and vice versa. Record the price and gas mileage of each. Assume you will drive a car for 10,000 miles each year for 10 years. Of those 10,000 miles, half will be city driving and half will be highway driving. Assume gas prices will be a constant $1.05 over those 10 years. Calculate the cost of operating each vehicle for the 10-year period. Just like appliances, the purchase price is included in the long-term cost of operation. For simplicity, do not include maintenance costs.

Data Sheet 17.1

Appliances, Lights:

Item	Minutes used	Kilowatt-hours	Purchase price	Annual energy usage	Annual energy cost	10-year cost (purchase price + 10 yrs. of energy cost)
100-watt incandescent Bulb						
40-watt fluorescent Bulb						
Refrigerator (home)						
Refrigerator 1 (store)						
Refrigerator 2 (store)						
Electric stove (home)						
Electric stove 1 (store)						
Electric stove 2 (store)						
Gas stove (home)						
Gas stove 1 (store)						
Gas stove 2 (store)						
Microwave						
Television						
CD player						
Computer						
Washer						
Dryer						
Automobile 1						
Automobile 2						

Data Sheet 17.2

Name _____

Section _____

Date _____

Water:

Number of pounds of water collected in 15 minutes	Amount of water lost per year	BTUs lost
	× 35,040 =	× 80 =

Transportation:

Mode of transportation	Miles traveled in 1 week	Cost
Automobile		
Foot		
Bicycle		
Public transportation		

Your Car:

A. Make and model:_____

B. Year: _____

C. Price: _____

D. Miles/gallon (city): _____mpg

E. Miles/gallon (highway): _____mpg

F. Gallons for city driving: _____gallons = 5,000 miles
(estimate 5,000 miles/year) mpg (from B above)

G. Gallons for highway driving: _____gallons = 5,000 miles
(estimate 5,000 miles/year) mpg (from E. above)

H. Cost for driving in city: $_____ = (_____ gallons for city driving, from F) × ($1.05/gallon)

I. Cost for driving on highways: $_____ = (_____gallons for highway driving, from G) × ($1.05/gallons)

J. Total cost of driving your car for one year: $_____ (H & I)

Opposite car:

A. Make and model:_____

B. Year: _____

C. Price: _____

D. Miles/gallon (city): _____mpg

E. Miles/gallon (highway): _____mpg

F. Gallons for city driving: _____gallons = 5,000 miles
(estimate 5,000 miles/year) mpg (from B above)

G. Gallons for highway driving: _____gallons = 5,000 miles
(estimate 5,000 miles/year) mpg (from E. above)

H. Cost for driving in city: _____ = (_____ gallons for city driving, from F) × ($1.05/gallon)

I. Cost for driving on highways: $_____ = (_____gallons for highway driving, from G) × ($1.05/gallons)

J. Total cost of driving your car for one year: $_____ (H & I)

VI. Questions

1. Were you more or less energy conscious than you thought before this lab? Explain.

2. Where were you most efficient in your energy usage? The least efficient?

3. Based on your comparisons, which were least expensive to operate, gas or electric appliances? Was it a large difference?

4. Based on how much gas you used in a week, what percentage of the purchase price of your car do you spend on gas each year?

5. Do you plan on changing your habits based on your results? If so, how. If not, why not?

I. Objectives

This chapter addresses the characteristics of renewable energies, the exchange of energy in the form of transfer of heat, and examples of extracting usable energy from the sun. After completing this exercise, the student will:

1. Understand the principles of transfer of heat energy.
2. Examine the storage of heat energy extracted from the sun.
3. Examine a method of transferring heat energy that has been stored.
4. Gain an awareness of the use of solar energy by humans.

II. Introduction

There are many renewable energy sources available for use on our planet. We can derive energy from the sun, wind, water, biomass, and through many forms of conservation. In developing energy resources, we can concentrate on a single source of available energy (i.e., solar energy in the desert), or we can utilize multiple sources of energy by integrating several systems (i.e., solar, wind, and biomass). The renewable energy resources that a country can develop depends on that country's location, the availability of various resources, and that country's economic condition.

In order to develop a strong energy base, we need to increase the availability and use of renewable resources. The most valuable and available renewable energy source we have is solar energy. Solar energy reaches the earth in various forms, one of which is heat. In order to use heat energy, however, we must transfer it from one form or place to another. There are three forms of heat transfer: conduction, convection, and radiation.

Water is a good example of a substance capable of transferring heat energy from one source to another. Water can be heated by a number of energy sources, such as a fire, a hot water heater, or, as described in this chapter, the sun. Heat energy is transferred from the heat source (i.e., the sun) to the liquid (i.e., the water) by **conduction,** and is then moved through the water by **convection.** We can quantify this transfer of heat energy by defining it in terms of thermal units. For example, one type of thermal unit, the **calorie,** is defined as the amount of internal energy added to or extracted from one gram of water, which will cause the internal energy to change one degree (+ or −) Celsius.

In this lab, we will use the sun's energy to heat water. We will be able to quantify changes in solar energy by measuring the change in water temperature caused by the energy from the sun. If we know the mass of the water and if we measure its change in temperature, we can determine the amount of energy added to the water. The equation below demonstrates the relationship between the energy gained or lost (in calories), the change in water temperature, and the mass of the water:

$$E = (t_f - t_i) * m$$

E = energy gained or lost
t_i = initial water temperature (°C)
t_f = final water temperature (°C)
m = mass of the water in grams

In the first part of this lab, we will heat several containers of water using the sun as the energy source. We will measure the initial temperature of the water (before exposing it to the sun) and the final temperature (after exposing it to the sun's heat energy). The change in temperature represents the amount of solar energy reaching the water and being absorbed by it. If the container is glass, the solar energy absorbed is the portion of the solar energy that is transmitted through the glass. Changing the container in some way (i.e., such as covering a part of the glass surface with aluminum, or painting the glass black, affects the solar energy transfer. When we use aluminum, we can either block the sun's rays, or concentrate them, depending upon how we position the container with respect to the sun. If we use a metal can, the metallic surface absorbs the heat energy, and transfers it by conduction through the metal to the water inside, where convection moves the heat through the water. If the source of energy is removed (i.e., if the sun goes down), then energy will move from the walls of the container out to the surroundings until the temperature of the container and its surroundings is the same.

Energy from the now-hot water can be removed from the water by a heat exchange mechanism. We will use a coiled copper pipe placed in the water. If cooler water flows through the pipe, the temperature of the water leaving the pipe will be higher than that entering the pipe, because the pipe will absorb heat from the hot water. Thus, the internal energy of the water in the pipe will have been increased. Likewise, the internal energy of the water in the container will have decreased.

III. Relevance: Competition for Nonrenewable Energy

The countries of the developed world use enormous amounts of energy. Indeed, these industrialized, highly technological societies could not exist without huge supplies of energy. The enormous growth in the consumption of energy by these societies is a major problem that we face. In fact, the increase in the use of energy in the United States is greater than the increase in its population. Even though developing nations have a much higher rate of population growth, they use much less energy per capita than the developed nations. It will be difficult for the developing countries to evolve without obtaining large increases in the availability of energy.

Thus, both the developed and developing nations need energy. Until recently, the largest sources of energy have been from nonrenewable resources (i.e., from fossil fuels), which are being rapidly depleted. Today, many people are involved in creating alternative methods of energy using renewable resources, and in conserving and making more efficient use of energy. Developed nations need to diversify their energy base and use various forms of energy in order to reduce their dependence on one energy source. Developing countries must expand production of their abundant, largely untapped, renewable energy resources. Both types of nations must begin to rely more upon renewable energy sources, and less on nonrenewables.

IV. Activity

In this lab, we will measure the temperature of the water in various containers before and after exposure to the sun, and differences between the tap water before it enters the coiled pipe, and after it exits the pipe. We will also manipulate the number of coils in the pipe to determine if these factors influence the amount of heat energy transferred from the heated water into the water in the pipe.

In measuring the energy (in calories) of the heat transferred, if the calculations result in a negative energy (i.e., if the final temperature is less than the initial temperature), this is fine, and simply indicates that energy left the system and was absorbed by some other system.

V. Procedure

Materials

(Per class)

> 3 one-gallon glass jars—marked at the 1000 ml level
>
> 1 one-gallon glass jar, painted black—marked at the 1000 ml level
>
> aluminum foil

(Per group)

> 1 of the above containers
>
> 1000 ml graduated cylinder—marked at the 400 ml level
>
> 1 lab thermometer
>
> 1 coil of copper (or brass) tubing (¼- to ⅜-inch diameter), with coil comprising 6–8 turns
>
> 1 coil of copper (or brass) tubing (¼- to ⅜-inch diameter), with coil comprising 3–4 turns
>
> 1 piece of rubber tubing that fits a cold water tap and the copper tubing
>
> 1 piece of rubber tubing that fits the copper tubing
>
> a place to expose the above containers to the sun at the same time

Method

In the early morning, before lab begins, selected students or the instructor will do steps 1 through 6. When class time arrives, the class will divide into four groups, and each will be assigned one container. Each group will work with and take readings from their one container.

1. Selected students, or the instructor, will set up the equipment as follows.
2. Preparation of the containers:
   a. A clear-glass jar will be used by Group **1**, later in class.
   b. Two glass jars will have half of their glass covered with aluminum foil by taping the foil over half of the wall of the jar (do **not** put the foil completely around the circumference of the jar; cover half of the container's surface from top to bottom, leaving the other half clear glass). These jars will be used by Groups **2** and **3** during class.
   c. The black-painted jar will be used by Group **4**, later in class.
3. Using the graduated cylinder, fill all the containers with 1000 ml of water, and record the amounts on Data Sheet 18.1.
4. Measure and record the initial temperatures (t_i) of the water in each jar and record on Data Sheet 18.1.
5. Put the jars outside at the beginning of the day. Arrange the four containers in full sun as follows:
   a. clear glass jar: in the sun.
   b. one aluminum-covered jar: put the foil-covered side opposite the sun so that the sun strikes the clear glass side.
   c. the other aluminum-covered jar: position the jar so that the sun strikes the foil-covered side first.
   d. black-painted jar: in the sun.
6. Measure the initial temperature (t_i) of all four containers and record in table 18.1.
7. Measure the final water temperature (t_f) of each jar at the start of class. Record on Data Sheet 18.1.

8. Place the 6–8 coil tubing in the heated water of your container ("heating container"). With the first rubber tubing, connect the coil to the cold water tap. Connect the second piece of rubber tubing to the other end of the coil, and put the free end of the tubing into the graduated cylinder ("receiving container"). Turn on the tap, and **very slowly** dribble water through the coil and into the receiving container until you get **400 ml** of water. If you run water through too fast, there will not be enough opportunity for heat exchange from the water in the heating container to the water in the pipe.

    a. Measure the temperature of the stored water in the heating container (t_i).
    b. Measure the temperature of the water in the receiving container (t_f).
    c. Record the above on Data Sheet 18.2.

9. Repeat step 8, but this time use the 3–4 coil pipe. Record and do calculations on Data Sheet 18.3.

10. Write your results on the board, and get the results for the other containers. Record them on the appropriate data sheets.

Data Sheet 18.1

Name _____

Section _____

Date _____

Group Number: _____

A. Record the temperature (t_i) and (t_f) of the water in the containers.

Table 18.1   Initial vs. Final Water Temperature of Storage Container

	Clear glass jar	Aluminum side away from sun	Aluminum side towards the sun	Black-painted jar
Volume of water (ml)	1000	1000	1000	1000
t_i				
t_f				

t_i = initial temperature of water

t_f = final temperature of water

B. Compute the solar energy added to the water:
   1. To calculate the mass of water: 1 ml = 1 gm water

      m = __1000__ ml water = __1000__ gm of water
   2. E = (t_f − t_i) * m

      = _____ calories

Data Sheet 18.2

Name _____

Section _____

Date _____

Group Number: _____

C. Record the temperature (t_i) and (t_f) of the water in the heating and receiving containers.

Table 18.2 6–8 Coil Pipe

	Clear glass jar	Aluminum side away from sun	Aluminum side towards the sun	Black-painted jar
Volume of water (ml)	400	400	400	400
Heating t_i				
Receiving t_f				

t_i = temperature of water in the heating container
t_f = temperature of water in the receiving container

D. Compute the energy transferred from the heated water to the water inside the 6–8 coil pipe.
   1. To calculate the mass of water: 1 ml = 1 gm water
      m = __400__ ml water = __400__ gm of water
   2. $E = (t_f - t_i) * m$
      = _____ calories

Data Sheet 18.3

Name _____

Section _____

Date _____

Group Number: _____

E. Record the temperature (t_i) and (t_f) of the water in the heating and receiving containers.

Table 18.3   3–4 Coil Pipe

	Clear glass jar	Aluminum side away from sun	Aluminum side towards the sun	Black-painted jar
Volume of water (ml)	400	400	400	400
Heating t_i				
Receiving t_f				

t_i = temperature of water in the heating container
t_f = temperature of water in the receiving container

F. Compute the energy transferred from the heated water to the water inside the 3–4 coil pipe.
1. To calculate the mass of water: 1 ml = 1 gm water
   m = __400__ ml water = __400__ gm of water
2. E = (t_f − t_i) * m
   = _____ calories

VI. Questions

1. In the first part of this activity, which container absorbed the most heat energy and reached the highest temperature? Which remained at the lowest temperature? Explain the differing temperatures of the containers.

2. Since the mass of water was the same in each container, which appears to be the most efficient container in extracting solar energy? Why?

3. In transferring heat energy from the heating container, through the coils, and into the receiving container, was there any difference in the amount of energy transferred by the different containers?

4. Which combination of container and number of coils allowed for the greatest transfer of energy from the storage container to the receiving container? Explain.

5. Based upon what you have learned in this lab, describe the type of solar water heater (the type of collector, the number of coils, and the water speed), that would be most efficient in heating water, and/or transferring the heat energy to another water source.

The Effectiveness of Insulation

I. Objectives

In this exercise about energy efficiency, the student will:

1. Test the insulating effectiveness of various kinds of materials.
2. Perform a qualitative comparison of insulating materials.
3. Understand the importance of correct insulation to increase energy efficiency.

II. Introduction

One of the most economical ways to reduce heat loss or gain in buildings is by using appropriate insulation in construction. It is also possible to install insulation in already constructed buildings to reduce heat loss or gain. Most insulating materials are rated as to their insulating value. The standard unit used to describe insulating value is an R-value, a material's ability to resist the flow of heat through it. The higher the R-value, the better the insulating ability. 1/R (the reciprocal of R) is a measure of the amount of heat energy in British thermal units (BTUs) that would pass through a piece of material 1 square foot in area in 1 hour when the temperature is 1° Fahrenheit higher on one side of the insulation than on the other.

$$\text{heat loss/gain in BTU per hour} = \frac{\text{ft}^2 \times \text{difference in temperature (}^\circ\text{F)}}{\text{R-value}}$$

A BTU is the amount of heat energy necessary to raise 1 pound of water 1° F and is equal to 252 calories. The following table lists typical R-values for several kinds of materials.

Material	R-value
No insulation	0
Single-pane glass	0.9
Double-pane glass	1.85
Triple-pane glass	2.8
1-inch wood	1–1.5
1-inch fiberglass batting	3.1–3.7
1-inch Styrofoam	5.5

III. Relevance: Increasing Energy Efficiency

One of the major themes in achieving sustainability is efficient use of resources. While we may not yet be ready to switch all of our energy sources to renewable forms, we can make much better use of the nonrenewable energy we do use. Our current use of energy is extremely inefficient. For example, we drive cars that actually get worse gas mileage than several years ago. Our homes, also, are often grossly inefficient energy users. During winter, many homes let in large amounts of cold air; during summer, air conditioners must work overtime to cool inefficient homes. Another major factor in inefficient buildings is the lack of adequate insulation. Older homes, especially, may lack roof and wall insulation, which results in drafty, cold buildings that never seem to get warm. Luckily, these problems are quite easily solved. Old homes can be remodeled, and modern types of insulation installed. New homes can be built that are super efficient, with insulation, double-paned windows, efficient appliances, and other energy-saving devices. If we are to preserve our nonrenewable sources of energy until we can replace them with renewable sources, we must find more efficient ways to use energy.

IV. Activity

You will use a simple apparatus to test and compare the relative insulating ability of several materials. One chamber of a wooden box contains a light bulb that, when turned on, will heat the air. Depending on the effectiveness of the insulating material between the chambers, the second chamber will be heated to some extent. The students will be able to compare the temperatures of the second chamber when various insulating materials are placed between the two chambers. The more effective the insulating material is, the less heated the second chamber will become.

V. Procedure

Materials

Per group:

Wooden heating apparatus with lid consisting of:

> 2 adjacent chambers, each 8 inches tall, 8 inches wide, and 18 inches long
> electrical outlet that holds a light bulb in one chamber of the box
> 75-watt light bulb

Thermometer inserted through lid of the second chamber by a standard 1 hole stopper inserted into the lid

Various insulating materials cut to a size (8 × 8 inch square) that fits between the two chambers

Examples of insulating materials:

> single-pane glass
> double-pane glass
> triple-pane glass
> 1-inch-thick wood
> 1-inch-thick fiberglass batting
> 1-inch-thick Styrofoam

Method

Use the apparatus provided to test the insulating ability of several materials in the following way.

1. Insert the thermometer in the top of chamber #2.
2. Place insulating material in the middle of the apparatus (see figure 19.1).
3. Turn on the light bulb.
4. Record the change in temperature every five minutes for thirty minutes.
5. Graph the data on Data Sheet 19.1. Use different colored pencils if you test more than one type of insulating material.
6. Share your results with the rest of the class.

Your instructor will probably divide the class into groups, with each group responsible for testing one or two different materials. In order to have a basis for comparison, at least one trial should be run with no insulating material in the apparatus. The other materials may be assigned by the instructor or chosen by the students.

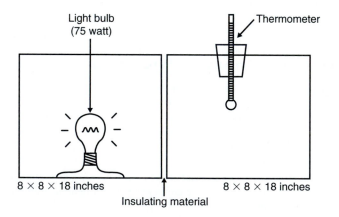

Light bulb
(75 watt)   Thermometer

8 × 8 × 18 inches   8 × 8 × 18 inches

Insulating material

Figure 19.1   Apparatus for measuring insulation efficiencies of different materials

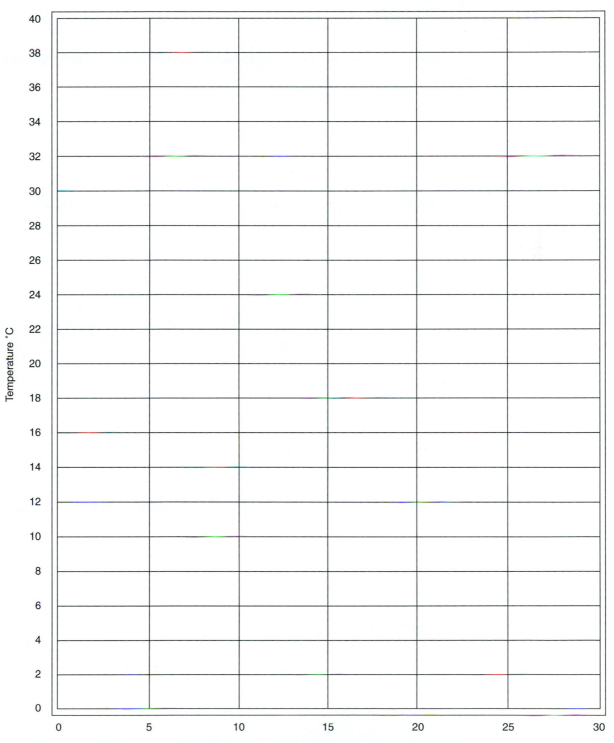

VI. Questions

1. Which of the materials you tested provided the most effective insulation?

2. If you had to choose the thinnest insulation to use, what characteristics would that insulation have to have in order to work the best?

3. Compare your results with published R-values for the material you tested. Are your results consistent with the R-values?

4. Which type of insulation is used in your home, apartment, or dorm? How could you find this information (without poking holes in your walls!)?

5. Based upon what you have learned in this lab, describe the type of insulation that would be most efficient in keeping a building cool in the summer and warm in the winter.

6. What, besides insulation, could you do to increase the energy efficiency of a house if you were going to design it for maximum sustainability?

EXERCISE 20
Personal Energy Consumption

I. Objectives

This exercise examines the amount of energy on individual uses. In this lab, the student will:

1. Calculate energy savings from reducing the size of a window or changing the kind of window.
2. Calculate energy loss from a dripping hot water faucet.
3. Examine the implications of an individual's lifestyle on energy consumption.

II. Introduction

Except for the small countries of Luxembourg, Bahrain, Qatar, and Oman, North America uses more energy per person than all other parts of the world. This is true because historically we have always had abundant, inexpensive energy in the form of wood, coal, and oil. Because we have had a large supply of inexpensive energy, there has been less interest in developing ways to use energy more efficiently.

There are several categories of personal energy consumption that we all have some control over. In our homes we can regulate heating and air conditioning, heating water, lighting, and the purchase and use of electrical appliances. We also can determine the methods of transportation we use, and how often we use them.

The units used in North America to measure quantities of energy are quite diverse. Quantities of heat energy are commonly given in British thermal units (BTUs), whereas electrical energy is usually measured in kilowatt-hours. Rather than try to convert all of the different units to the metric equivalent, we will use the standard units used in ordinary commerce.

BTUs in various amounts of fuel

1 gallon of fuel oil: 145,000

1 cubic foot of gas: 1031

1 kilowatt-hour of electricity: 3412

1 ton of coal: 25,000,000

1 cord of wood: 20,000,000

1 gallon of gasoline: 125,000

III. Relevance: Living More Efficiently

The purpose of achieving a sustainable lifestyle is not to make people uncomfortable, or to send them back to "living in trees." Sustainability merely means living within limits that ensure future availability of resources for all species, not just the current generation of humans. Each human has an impact on the environment, but some have a greater impact than others, depending on the amount of energy and resources each person uses. Throughout the day, the amount of electricity, water, gasoline, and lighting that a person uses can be a personal decision. If each person were to turn off unnecessary lights, turn the air conditioner to a more efficient setting, use a bike instead of a car, plant native plants instead of water intensive exotics, and so on, the cumulative impact of society on the environment would greatly decrease. Becoming more aware of the energy and resources one uses, and then acting to decrease what is used, may be hard at first, but can become an ingrained habit, just like brushing your teeth. A sustainable lifestyle starts with the individual, making choices that decrease his or her impact on the environment.

IV. Activity

This lab examines the types of energy students use (or misuse!) during a normal day. The energy choices people make to heat and cool their living areas, to light their houses, in their electrical appliances, and in the types of transportation they use all put an enormous burden on the environment. Before anyone can make a wise decision on how to increase his or her personal efficiency, the first step usually involves an assessment of how much energy that person uses. Students in this lab will, either individually or in groups, measure the amount of energy they use for heating water, lighting their homes, driving, and using electrical appliances. Students will also calculate how much energy would be lost by inefficient appliances, as well as energy that would be saved by switching to efficient appliances.

V. Procedure

Note: If there are time constraints that do not allow the student to do all of the following activities, the class can divide into groups. Each group will then be responsible for gathering data on: (1) lights, (2) major appliances, (3) heating water, (4) transportation, and (5) heating and

161

air conditioning. In class, the groups can then pool their data, and fill in the data sheet. An alternative idea can be for the students to gather information for a shorter time period (e.g., 2 hours), and then extrapolate this information to provide data on an entire 24-hour period. For example, if lights are left on for 1 out of 2 hours, then the data could be extrapolated to mean that, during periods of darkness, the light is on 50 percent of the time.

Fill in the following information in the appropriate tables of Data Sheets 20.1 and 20.2.

Heating and Air Conditioning

One of the major ways that energy leaves or enters buildings is through windows. A single-pane window has an R-value of 0.9. 1/R is equal to the number of BTUs that would pass through a 1-square-foot surface in 1 hour if the difference in temperature on opposite sides of the surface is 1° F. Therefore, we can calculate heat loss or gain through a window by using the following formula:

$$\text{heat loss/gain in BTU per hour} = \frac{\text{ft}^2 \times \text{difference in temp. (°F)}}{\text{R-value}}$$

Record all results from steps 1–8 on Data Sheet 20.1.

1. Choose a single-pane window in your classroom or home and measure its surface area.
2. Measure the difference in temperature between the inside and the outside of the window.
3. Calculate the rate of heat transfer through the window by using the formula above.
4. Assume the same temperature conditions existed for an entire year. Calculate the heat transfer.
5. Double-pane windows have an R-value of 1.85. Triple-pane windows have an R-value of 2.8. Calculate the effect on the rate of heat transfer if the single-pane window were replaced with double- or triple-pane windows.
6. Obtain the R-value for special low-emission glass and calculate heat transfer for these windows.
7. Calculate the effect a 5° F decrease in the temperature on the inside of the window will have on the rate of heat transfer.
8. Calculate the rate of heat transfer if the size of the window were reduced by 50 percent.

Heating Water

Water resists changes in temperatures. In other words, it takes a lot of heat energy to make a small change in the temperature of water. Therefore, water heaters are quite expensive to operate. It takes 1 BTU of heat energy to raise the temperature of 1 pound of water 1° F. Record all your results on Data Sheet 20.1.

1. Turn on a water faucet so that it is leaking at a rate of about one drip per second.
2. Capture this water for a period of fifteen minutes.
3. Weigh the amount of water you have collected (in pounds).
4. Multiply this number by 35,040 to find the amount of water that would be lost in one year.
5. Assume that water entering the water heater enters at 40° F and leaves the heater at 120° F.
6. Calculate the number of BTUs of heat energy that would be lost in one year if the leak were not fixed.

$$\text{Number of BTUs/year} = \text{pounds of water in 15 min} \times 35,040 \times 80 =$$

7. Calculate the number of kilowatt-hours of electricity it would take to produce this much heat.

$$\text{kilowatt-hours} = \frac{\text{BTU}}{3412}$$

8. Calculate the cost associated with this water loss by multiplying the number of kilowatt-hours by the cost of a kilowatt-hour of electricity. (Obtain the cost by looking at your electrical bill.)

Lighting

Lighting is something that we take for granted. We usually simply flip a switch and it is instantly there. However, what does it cost in energy to provide the light we use and does it make a difference what kind of light we use?

1. In a dark room, hold a light meter exactly 10 feet from a 40-watt incandescent light bulb.
2. Record the reading on the light meter.
3. Use the light meter to measure the amount of light coming from a 40-watt fluorescent light bulb.
4. Since both bulbs have the same wattage, they use the same amount of electrical energy in the production of light. Which of the bulbs is more efficient in providing light?
5. Approximately how much more efficient is this bulb?
6. Record your results on Data Sheet 20.1.

Transportation

North Americans look at freedom of movement as a right. We drive and fly more than any other people in the world. In some urban areas, trains provide efficient ways to travel about the city, reducing traffic jams and air pollution.

1. For one week, keep a log of all the miles you travel by the following methods:
   a. Foot
   b. Bicycle
   c. Automobile
   d. Train or other rapid transit
   e. Plane
   f. Other

2. Approximately what percentage of each of these was done just for fun?
   a. Foot
   b. Bicycle
   c. Automobile
   d. Train or other rapid transit
   e. Plane
   f. Other

3. Record your results on Data Sheet 20.2.

Electrical Appliances

Electrical appliances are very convenient. They allow us to do things quickly and relieve us of distasteful or tedious tasks. How much energy do you use as a result of such devices? It is important to recognize that the total energy cost of an appliance also includes the energy necessary to manufacture, distribute, and merchandise the item. However, it might be instructive to determine how much energy is consumed by the use of the various electrical appliances. You can find the wattage of an electrical appliance on a label on the appliance:

1. Keep a log of all the electrical appliances you use in a one-week period. List the appliance and the number of minutes it was used per week.
2. Record the wattage of the appliance from the label on it. If the label does not show the wattage but gives volts and amperes, you can calculate wattage as follows:

$$Watts = Volts \times Amperes$$

3. If you know the wattage and the number of minutes it was used, you can calculate the number of kilowatt-hours of energy used (see the equation below).

$$Kilowatt\text{-}hours\ used = \frac{(watt\ rating)(total\ minutes\ used/60)}{1000}$$

4. If 1 kilowatt-hour is equal to 3413 BTUs, how does energy consumption by using electrical appliances compare to energy consumption by automobiles or for home heating and cooling?
5. Record all results on Data Sheet 20.2.

Data Sheet 20.1:
Personal Energy Consumption

Name _____

Section _____

Date _____

Heating and Air Conditioning

Window	Surface area ft^2	Inside temperature °F	Outside temperature °F	Temperature difference	R-value	Heat loss or gain/ hour (BTU)	Heat loss or gain/ year (BTU)
Single-pane					0.9		
Single-pane (inside temperature 5° F cooler)					0.9		
Single-pane (surface area ÷ 2)					0.9		
Double-pane					1.85		
Triple-pane					2.8		
Low-emission glass							

The rate of heat loss/gain/yr if there is a 50% decrease in the outside temperature, with a single-pane glass: _____
The rate of heat loss/gain/yr a single-pane glass window were reduced 50% in size: _____

Heating Water

Number of pounds of water collected in 15 minutes	Amount of water lost per year	BTUs needed to heat water lost
	× 35,040 =	× 80 =

Kilowatt-hours of electricity needed to heat the water lost: _____
Cost of the kilowatt-hours of electricity used: _____

Lighting

	40-watt incandescent	40-watt fluorescent
Light meter reading		

How much more efficient is the fluorescent bulb? _____

Data Sheet 20.2

Name _____

Section _____

Date _____

Transportation

1. For one week, keep a log of all the miles you travel by the following methods:
    a. Foot _____
    b. Bicycle _____
    c. Automobile _____
    d. Train or other rapid transit _____
    e. Plane _____
    f. Other _____

2. Approximately what percent of each of these was done just for fun?
    a. Foot _____
    b. Bicycle _____
    c. Automobile _____
    d. Train or other rapid transit _____
    e. Plane _____
    f. Other _____

Electrical Appliances

Appliance	Wattage	Minutes used	Kilowatt-hours used
Microwave oven			
Electric stove			
Hair dryer			
Stereo system			
Home computer			
Television			
Dishwasher			
Garbage disposal			
Electric shaver			
Space heater			
Electric fan			
Washer			
Electric blanket			
Vacuum cleaner			
Electric clothes dryer			

VI. Questions

1. How much energy was lost in your home from inefficiency (e.g., dripping water faucets, lights left on when not needed, unnecessary automobile trips, appliances left on when not being used)?

2. Where were you the most wasteful? Where were you the most energy efficient? Why?

3. How can you improve your energy efficiency? How hard do you think that would be?

4. What alternative forms of energy can you use to do the same tasks in your home that currently use conventional forms of energy?

Field Trip Suggestions

1. Visit a power plant: nuclear, coal-fired, wood-fired, or hydroelectric. Determine the quantity of electrical energy produced, the size of the service area, if electricity is sold to other utilities, and whether the facility is used primarily for the base-load or peak-load requirements of the utility. Describe how the utility minimizes its negative ecological impacts.
2. Visit your school's power plant. Determine the amount of energy used per year, and calculate energy use per student per day. What steps has your school taken to reduce energy consumption? Why were they taken?
3. Visit a coal mine, oil field, or gas field. Describe how the company minimizes its negative ecological impacts.
4. Visit an energy information center. (Most power companies provide such services.) Describe five changes you would make and how much energy you would save. What would this mean in monetary terms?
5. Visit a pipeline or powerline right-of-way. Describe three ways the vegetation in the right-of-way differs from the adjacent, less disturbed land.
6. Visit an oil refinery. Describe how the company minimizes its negative ecological impacts. Determine where the company sells its product.
7. Visit a nuclear facility (hospital, X-ray installation, or nuclear power plant). List ten steps taken to assure safety.

Alternative Learning Activities

1. Invite a power company executive to talk to the class.
2. Invite the school's physical plant director to talk to the class about the energy requirements of the school and the costs involved.
3. Trace the path of oil from production facility to gasoline station by interviewing or writing letters to people and asking where they purchase their product (i.e., gas station gets its gas from a distributor, who gets it from a wholesaler, refiner, oil pipeline company, etc.).
4. Trace the path of coal to a power plant, uranium to a nuclear power plant, natural gas to a home.
5. Draw up a list of ten ways you could reduce energy expenditures. Implement your list.
6. Within your class conduct a contest to see who can reduce energy expenditures the most.
7. Collect all the wastepaper in a particular building or area of your campus and determine how much energy this represents in terms of calories of heat energy.
8. Visit a local hospital to learn how it generates low-level radioactive waste. How does it dispose of this waste? What does it cost?

PART 5

Lifestyle Choices

The ultimate way to achieve sustainability depends on the individual. Laws may be passed, regulations may be enforced, and the market place may place different values on alternate methods of pollution control, but it is ultimately the individual, and his or her choices, that will determine the future quality of life for people around the world. As more and more people become concerned and educated about environmental issues, they are changing their lifestyles and actions to decrease their impact on the environment. People can maintain (or even enhance) their quality of life, while at the same time protecting environmental quality. A little bit of thought, and small changes in everyday actions, will do a lot towards ensuring a sustainable environment and lifestyle. If everybody made an effort, the environment could be protected without undue sacrifice.

Part 5 is designed to enhance the individual's awareness of how common actions can impact the environment. Exercise 21 focuses on techniques that describe the public's perception of an environmental issue. Exercise 22 takes the students to a shopping mall to see how current urban planning impacts the environment, and how alternate methods may decrease this impact. Exercise 23 is designed to increase environmental awareness via a trip to the landfill, a place that contains the end result of a very consumptive lifestyle. Finally, exercise 24 uses formal research to help the student address an issue of particular interest to that student.

An Environmental Survey

I. Objectives

Following completion of this exercise, students will:

1. Construct, administer, and analyze a survey.
2. Understand the critical role surveys can play in environmental planning and decision making.
3. Understand the survey's use as a research method, its strengths, weaknesses, components, rules governing use, and design.

II. Introduction

A survey is a systematized, critical inspection or observation of an area, population, or behavior, which provides information (data) for analysis.

A survey is simply a way to obtain detailed information about a specific area, subject, or group. Once collected, this information can then be processed to determine relationships and correlations, which may be valuable in understanding and solving problems.

For example, many regions in the United States plan for continual access to safe drinking water. Often, however, unexpected developments occur, such as the explosive growth of the dairy industry in many areas. The runoff of organically polluted water from dairy and other agricultural activities upstream occasionally causes an algal bloom in lakes, which causes taste and odor problems in the drinking water. There is really nothing wrong with the water, but because of the odor and taste left by the algae, lake water often has a bad reputation. The problem would be solved merely by providing funding for installing additional treatment and filtration equipment.

A survey could be administered to determine how knowledgeable people in an area are about their drinking water. The **population** would consist of all citizens in the communities that get their water from the lake. The **sample** would consist of all people actually polled, since it is unrealistic to poll every single person (there may be millions of people in the community). **Variables** are measurable characteristics that include age, sex, social status, marital status, income level, education, daily/weekly access to news sources, types of media sources, cost of water, home ownership verses apartment rental, and the willingness to support a tax increase to fund the necessary improvements to ensure better water quality.

One possible statistic from the above mentioned study might be that 63 percent of those polled would pay more for better quality water. Valuable associations between certain variables could be found, which could help decision makers determine the best methods to improve people's perceptions about their drinking water. For example, age might be found to influence a person's willingness to pay more for water. Thus, different educational approaches could be taken that are targeted towards various age groups.

The survey as a research method:

Surveys are conducted in several ways. One may choose to interview people by mail, in person, or over the telephone. There are advantages and disadvantages to each type of survey. For example, mail surveys can decrease interviewer bias because the interviewer's facial expressions or body language cannot affect the respondent's answer to a question. One disadvantage of mail surveys is the fact that they do not have a high response rate. For this lab, we will construct a written questionnaire that the lab students will administer by telephone, which decreases interviewer bias (because the respondent cannot see the interviewer's expressions) but has a higher response rate than mail surveys.

The adequacy of a survey is discussed in terms of its validity and reliability. If the same survey were readministered under similar conditions, and the results were the same, the survey would be considered **reliable**—it yields the same results each time it is done. If a survey measures what it intended to, the survey is considered **valid.** For example, if survey results showed that people were willing to pay higher taxes for improved water quality, and the people in the community then voted to pay higher taxes, the survey would be a valid one.

Survey research has several strengths. It allows the researcher to gather large amounts of data on large numbers of people and is accurate within a predefined range. Written questionnaires reduce interviewer bias and guarantee uniform question presentation.

Surveys also have several weaknesses: (1) they may be costly in terms of time and money; (2) they are based on the assumptions that respondents will give truthful answers, (3) the sampled population represents the larger population from which the sample is taken; and (4) reliability and validity are difficult to check. The value of the survey depends on how detailed the design is, and on the quality of the data gathered.

III. Relevance: Everybody Has an Opinion

Surveys, and the data they generate, are important in environmental decision making. Land-use planning, political decisions, regulation of toxic waste disposal, creation of national parks and recreation areas, and the use of public funds for mass transit all depend on information provided by surveys.

In democratic societies, laws are often the result of public opinion. Public opinion is a function of many influences, including knowledge of the issue at hand. Optimally, this knowledge results from objective scientific discovery. Increasingly, however, public opinion is formed by media sources, which may not be fully informed, and are even influenced by political, economic, or other vested interests.

One type of survey, termed a **public opinion poll,** can determine how the public feels about environmental issues. In turn, these polls may influence political decisions about the environment. Given the critical nature of environmental decisions to you and to future generations, it is important for you to be able to understand surveys and their significance.

IV. Activity

For purposes of this lab exercise, we will perform a phone interview survey. For this activity, a questionnaire must be constructed to collect the data.

Constructing a questionnaire is both an art and a science and usually involves five steps:

1. Specifying the research objectives
2. Selecting the type of questionnaire
3. Writing the first draft
4. Pretesting the questionnaire
5. Revising the questionnaire

Questionnaires may also be distinguished by the structuring of the questions and responses. In **closed-ended questions,** the respondent is asked to select his or her answer from a list of alternatives.

EXAMPLE: "What do you think is the most serious problem facing the United States today?"

A)   Racial relations
B)   Poverty
C)   Inflation
D)   Environmental degradation
E)   Unemployment
F)   Drugs
G)   Other (please specify) _____

The response categories should include all major alternatives, where the answer falls in only one category. For example, a person's age cannot be contained in both choices, as in (a) 36–40 years and (b) 41–45 years. Closed-ended questions are useful when the researcher knows enough about the topic to select appropriate responses.

Open-ended questions let the respondent provide the answer to the question without being given any alternatives. These types of questions tend to be more varied and can be harder to analyze. For example: "What is the most serious problem facing the United States today?" can have an endless number of responses, so that the analysis of this question would involve lumping similar responses into broad categories.

Rules for Use

Wording the Question—The phrasing of questions can be a very difficult process. Some of the more important guidelines for constructing questionnaire items are listed below:

1. Questions should not be complex or double-barreled. Asking for a yes or no answer to a complex question involving a number of parts can be confusing or misleading.

Poor	**Improved**
Do you support environmental controls and clean air and water, or not?	Do you support environmental controls? Do you support increased taxes to improve air quality?

2. Questions should not contain biased or emotionally loaded terms, which can influence respondent behavior, skew data, and affect the survey's validity.

Poor	**Improved**
Do you favor or oppose the proposed child-killing abortion bill?	Do you favor HB 211, which forbids voluntary termination of a pregnancy during the first trimester?

3. Questions should not seek technical information that the respondent cannot answer.

Poor

Which treatment technology do you think the City of Waco should install to solve its water taste and smell problems: ozone diffusion, reverse osmosis filtration, electrochemical flocculation, or granular activated carbon charcoal filtration?

4. Questions should not be stated negatively, which can confuse or mislead a respondent.

Poor	**Improved**
Should marijuana not be legalized?	Should marijuana be legalized?

Question Sequencing—The order in which topics are addressed in the questionnaire, as well as question type, influences responses to the questions. If a closed-ended question precedes an open-ended one relating to the same topic, the information contained in the closed-ended question may influence the response to the open-ended one. Accordingly, open-ended questions normally go before closed-ended ones.

Sample Selection—It is often impractical to sample every individual in a population for a survey. However, the validity of the survey requires that the sample surveyed represents the larger population from which the sample was selected. The best way to accurately reflect the characteristics of a population is through **random sampling,** where every person in the population has an equal probability of being included in the sample.

A Sample Survey

The local city planning department wanted to know more about citizen use of municipal parks in order to provide the best recreational opportunities for the broadest range of people.

Research questions for this study might include such information as:

- Do citizens use parks?

- Who uses the parks?

- How often?

- Which parks?

- What activities are pursued there?

- If not using now, what would encourage use?

- Would citizens support a tax increase to pay for better parks?

Recording Results

Each person answering the survey has his or her own answer sheet, either filled out by an interviewer (as in this exercise) or by himself or herself. Once all the people in the sample have been surveyed, a **tally sheet** is constructed to display answers from all the individual answer sheets. One easy way to do this is to record each individual's responses from the answer sheets on a separate line of notebook paper. With a ruler or straight edge, a tally sheet similar to the following one displaying results of our hypothetical sample survey (where three people have answered the questions) can be constructed. The total number of each response for each question is easily counted. Responses such as number of children and age can be shown as an average (e.g., average age = 30 + 25 + 50 = 105/3 = <u>35</u>) while other responses are best shown as a percentage (e.g., question #5 = 67% ($\frac{2}{3}$) agree, while 33% ($\frac{1}{3}$) disagree with the question).

| | Variables | | | | | Independent variables | | | |
Person	Ques. 1	Ques. 2	Ques. 3	Ques. 4	Ques. 5	Sex	No. of kids	Marital status	Age
1	often	lake	picnic	more	agree	F	0	S	30
2	freq.	lake	sports	more	agree	M	2	M	25
3, etc.	often	ocean	relax	less	disag	F	3	D	50

You must record your data in such a way as to allow its compilation with the data gathered by the rest of the class, because we will tabulate all data from every student for each question.

Please come to class with your data in tables. *Hint:* you can construct a table, where the first column consists of all people you interviewed (1 per row), and each subsequent column contains the answers to one of the questions, so that if you asked 10 questions to 10 people, you would have an 11 row ("header" row containing the question numbers + 10 respondents, 1/row), 11 column (respondent column + 1 answer/column) table. You can then summarize, at the bottom of the table, the number of people that answered each question as (a), (b), etc. (see example on next page).

Example summary table (3 respondents, 4 questions):

Respondent	Question 1	Question 2	Question 3	Question 4
1	yes	a	a	c
2	yes	b	b	b
3	no	a	b	a
Total	2 yes	2 a	1 a	1 a
	1 no	1 b	2 b	1 b
		0 c	0 c	1 c

V. Procedure

Materials

    telephone book
    random numbers table (see Appendix A)
    calculator
    prepared answer sheets

Methods

The entire class will choose a topic and design a survey in class. During the week, each student will survey ten people and tally his or her results in a table.

1. The class will choose a survey topic dealing with public awareness of an environmental issue. This can cover a wide range of topics such as water taste and odor problems, air pollution, city planning, environmental attitudes, ethics, public land use, or any specific local issues. Please come to class with some idea of a suitable topic and a recommendation for a survey. Be creative!

2. Using the suggestions in the Activity section, design your questionnaire. Write all potential survey questions on the board. Don't worry about the form, get the ideas down first. Then, work on the construction of the questions to avoid the mistakes discussed earlier. Select the best 10–15 questions for the final survey. Once the classes' questions are finalized, write them out neatly on the board for each student to write down.

3. Each student should make out ten answer sheets to record the responses from each person interviewed (1 person/sheet).

4. Before doing the telephone survey, use the random numbers table in Appendix A to select a page number in your student directory, since this survey is limited to fellow student's beliefs. Think of a number between one and ten. Now, using the random numbers table in the appendix, start from the top, bottom, or side and count across, up, or down to the number representing the number you had in mind. Let's say you chose seven. Starting at the top left column and counting down, the seventh number could be 60. Write this number down. Now count down another seven numbers to 21, and so on until you have ten numbers written down. In the local phone book, use the first of the ten numbers you picked, turn to any random page number, and count down until you get to your first number (e.g., the 60th phone number listed). That number will be the one you call. Call the person and conduct your survey. Record your results on the answer sheet as the person is responding to the questions. Repeat until you have interviewed ten people. Some people will be cooperative, but others may be "less than cordial"; in any case, be courteous and persistent. If the person does not want to be surveyed, select another respondent using the random numbers table. Identify yourself and say why you are calling. Thank the person for his or her time when you are through.

5. Summarize the results of your survey in a table, as described in the Activity section.

6. The next week in class, tally the responses for all students' results for all questions on the chalkboard. Make a copy of this for yourself on notebook paper.

VI. Questions

1. Look at the classes' total responses. Were there any patterns or interesting responses to the questions?

2. How would you improve the survey your class performed?

3. Were people responsive to your survey or did they show a lack of concern? Explain.

4. Would you say the sampled population is knowledgeable about environmental issues?

5. How could an environmental survey be used in community planning?

6. Were your results different from what you expected before you conducted the survey?

7. Where do you think most people get the information they use to formulate their opinions and attitudes about environmental issues?

Land-Use Planning: A Shopping Center

I. Objectives

After surveying people at a local shopping mall, the student will:

1. Understand that land-use decisions may not always be carefully thought through.
2. Recognize that most types of land use may have unwanted or negative environmental results.
3. Understand the relationship between the development of shopping centers and the decline of central business districts.

II. Introduction

Shopping centers are frequently constructed with inadequate total planning. Communities are quick to see the potential positive economic spinoff of a shopping center being developed, but tend to downplay potential negative impacts of such development. Environmental considerations are not always properly addressed, and often surface later as problems. To understand what environmental problems may arise from the construction of a shopping center, it is necessary to analyze an existing or proposed facility.

III. Relevance: Planning for Sustainability

Economic development is often seen as a desirable aspect of community growth. Community projects are built with the assurance that jobs will be created, and the community will prosper. Unfortunately, the negative aspects, both environmental and social, are often ignored. For example, proponents of building new highways in rural areas state that people will be able to travel faster from one area to the next, that tax dollars are well spent on such facilities, and that anti-highway people are anti-progress. Such arguments do not address the negative impacts of highways, however. Rural communities are often divided and destroyed by the increased development that results from greater access due to the highway. Irreplaceable farmland is lost to concrete and buildings. Habitat is lost, resulting in negative impacts on many species. People that moved to the country to escape the pressures of an urbanized setting may find that the quality of life diminishes with increased development. This is not to say that development is bad, but that future construction projects need to carefully take into account the needs and desires of the community. Communities can increase their economic well-being without sacrificing quality of life, but it will take careful prior planning and consideration.

IV. Activity

In this lab, you will take a field trip to a local shopping center (mall) and interview customers, using the questions described in the Procedure section (below). According to the time allotted for this lab, you may either go to the mall as a class, or work together in groups and interview respondents during the week, and bring your results in to be tallied in the next class. To record respondent's answers to the survey, each person answering the survey will have his or her own answer sheet, which will be filled out by an interviewer. Once all the people in the sample have been surveyed, construct a tally sheet (as described in exercise 21) of all respondents' answers. This tally sheet can be used in the next lab class to construct a tally sheet of the entire class, so that conclusions regarding the survey responses can be made.

V. Procedure

The week BEFORE this lab:

1. Decide whether the whole class will go to the mall and interview respondents, or whether students will divide up into groups, survey people, and report their results during the next lab class.

Conducting the survey:

2. Use the following questions as a basis for your survey. Add other questions that are pertinent to your situation.
3. When surveying people, be polite, and ask them if they would be willing to take a few minutes to answer some questions. Regardless of their answer, thank them and wish them a good day.
4. Sample questions:
   a. Did you live in this area before the shopping center was constructed?
   b. If so, what was the area like before the shopping center was built?
   c. What problems do you notice that are associated with the shopping center (noise, traffic, etc.)?
   d. Is the shopping center's interior aesthetically pleasing to you?
   e. Is the shopping center's exterior aesthetically pleasing to you?

    f.   Do you feel crime in your neighborhood has changed with the construction of the shopping center?

    g.   Were you consulted in any way before the shopping center was built?

    h.   If so, in what way and by whom?

5. After the survey, construct a tally sheet with all answers, and bring this sheet to class.

Analysis of the shopping center:

6. Sometime before the next lab class, return to the shopping center, and look around. Answer the following questions, and be prepared to discuss them in the next class.

    a.   Estimate the total area of impervious (e.g., concrete surfaces, parking lots, buildings, etc.) surfaces where water cannot seep into the ground.

    b.   Estimate the number and average height of trees in the shopping center.

    c.   How are the trees planted? In boxes, in beds, or in areas filled in with wood chips?

    d.   Estimate the percentage of the parking lot that is filled with cars.

    e.   Where does all the water that falls onto the shopping center drain into?

    f.   Is the area aesthetically pleasing to you? Why or why not?

g.   Were the roads leading to the mall there before the construction of the mall? Were additional roads built to accommodate the additional traffic? Were existing roads widened?

h.   What types of stores are in the mall? How many stores (total) are there?

i.   Is there evidence that the mall was constructed where a farm, residential area, or open fields used to be?

j.   Is there evidence that large trees were removed for the shopping center? If so, how could they have been incorporated into the design of the mall, rather than being removed?

k.   What type of wildlife is present at the mall?

VI. Questions

1. Did your respondents seem to feel positively about the mall?

2. What were some of the major problems respondents cited about having the mall in their community?

3. What environmental problems did you see that were associated with the mall (e.g., water runoff, aesthetics, wildlife, loss of habitat, traffic)?

4. What functions does this mall play in this community?

5. How can the negative environmental problems associated with malls be minimized?

6. Are several small shopping centers better than one large one? Why or why not?

7. If you were designing a mall, what important factors would you incorporate that weren't included in this mall?

EXERCISE 23

Environmental Awareness and Lifestyle

I. Objectives

After a field trip to the local landfill, the students will:

1. Be aware of how their everyday actions affect the local environment.
2. Be familiar with changes they can make in their lifestyles to improve the protection and conservation of resources.
3. Understand the function and operation of a municipal landfill.
4. Understand issues of solid waste disposal.

II. Introduction

Throughout this course, we have discussed many of the components that make up the world we live in and how they interact with each other. You now know that we are a part of nature and not apart from nature. We are still a long way from achieving the sustainable society that we need in order to keep a certain degree of quality of life on this planet. To achieve this, everyone will have to make changes in his or her personal lifestyle. However, working as an individual is not enough; one must also work within the decision-making infrastructures in order to achieve a sustainable society.

Sometimes environmental problems seem so overwhelming that one person's actions seem hopeless. However, there are many things that an individual can do to help the environment. The home is a good place to start. Water, electricity, and gas may all be conserved by retailoring personal use habits to use only what is necessary. Entrepreneurs are constantly developing and marketing new technologies to increase the efficiency of household appliances, such as low-flow showerheads.

On the community level, both awareness and action are important. Being aware of how your community functions, and problems associated with its functioning, are first steps in solving environmental problems. You must understand the dynamics of community utilities such as power production, water and sewage treatment, and solid waste disposal before you can start to solve environmental problems associated with them.

To effect changes, you must also understand the political and economic infrastructure of your local community. The interplay of power between elected officials and the business community cannot be underestimated when addressing environmental issues. A major part of solving problems is dealing with the people involved with the issue, and not just the scientific aspects of the issue. Until recently, short-term profit has superseded long-term environmental cost. Hopefully, people will realize that future costs must be built into present-day consumption in order to realize the full cost of unsustainable activities.

Train yourself to be attuned to environmental problems by observing what goes on in your community. Pay attention to local and national media coverage. Do not passively accept things as they are, just because they have always been done that way. Learn to be objective when viewing the world around you.

Once you understand the problem, the next step is to become involved. Attend meetings, join groups, form committees, and write to elected officials and newspapers. Actively participating in your community will allow your voice to be heard in helping preserve the environment for future generations.

III. Relevance: What Do We Do with All This Trash?

The disposal of solid waste is an increasing problem around the world. Each year, Americans dispose of 125 million tons of solid waste, approximately 2.6 pounds per person per day. About 15 percent of that waste is recycled, leaving 110 million tons to be put into landfills, incinerated, or dealt with in some other way. Currently, landfills are the most cost effective way of dealing with solid waste disposal. Through **tipping fees** (the fee required to enter and dump into the landfill; typically $25 to $150 per ton), landfills generate 45 billion dollars in revenue and employ 3 million people per year.

There is a downside to landfills, however. No one knows how long it takes solid waste to completely decompose. Landfills must constantly pump out leachate (water contaminated by garbage) to ensure it does not enter the drinking water. Once the pits reach capacity and are filled, landfills must be monitored for 30 years to ensure that leachate does not escape. In addition, landfills take up space. As human populations grow, and the landscape becomes more crowded, landfills will have to be built closer to people's homes. Often, people want garbage disposed of, but not near where they live. This means we either find new forms of solid waste disposal, reduce the amount of waste going into landfills, or get used to the idea of living next to them. There are several ways an individual can decrease the amount of solid wastes produced in the home. Glass, aluminum, paper, and plastics can be recycled. Organic wastes (i.e., yard clippings and inedible food) can be composted and turned into high-quality fertilizer. Gardening and wise shopping can greatly reduce the amount of packaged goods consumed in your home, and fresh foods tend to be healthier. Reusable shopping bags and containers can decrease the use of natural resources. By limiting the use of your car, through car pooling, using public transportation, and reducing the number of trips you take, you can reduce energy use and pollution. For short trips, walking or riding a bicycle is healthier, cleaner, and more energy efficient.

181

IV. Activity

Just as the trip to your local sewage treatment facility was designed to show you that the waste does not just disappear when you flush the toilet, this field trip to your local landfill demonstrates that what you throw into your trash can doesn't magically disappear on trash collection day.

Before you go to the landfill, think about the following questions, and be prepared to address them as you go through the tour.

1. What were the major environmental and legal issues involved with placement of the landfill?
2. What is the yearly cost to operate the landfill? How many tons of garbage end up here per day? Per year?
3. What type of garbage (i.e., paper, lawn clippings, plastic) takes up the most room? What is the proportion of different types of garbage at the landfill?
4. Describe how a landfill is constructed. How long will it be before your facility reaches capacity?
5. How is leachate handled in your landfill?
6. Who inspects and regulates landfills?
7. Are there efforts to recycle materials in the landfill?
8. What is the procedure for receiving and covering garbage?
9. Are individuals allowed to dump their trash, or is dumping limited to professional trash haulers?

V. Procedure

Your class will take a field trip to the local landfill. You will need to wear appropriate clothing for the tour, have your lab manual and questions, writing supplies, and a clipboard or other backing on which to draw and write.

Before the field trip, write the questions (above) down, and try to answer them. Also, before the trip, think of additional questions that have local importance, write them down, and ask them during the tour. While you are at the landfill, pay close attention to the characteristics of the landfill, the job requirements of the workers, the machinery used, and the local wildlife. During the tour, ask the guide questions you do not have answers to. Remember, it is up to you to ask the guide questions; he or she will not be able to predict what you will ask.

VI. Questions

1. What was the most interesting part of your field trip?

2. Do you see any recyclable materials in the landfill?

3. Do you see any shortcomings in the treatment of the garbage that enters the facility?

4. What happens if toxic wastes are accidentally dumped into the landfill? For example, what may happen to the wildlife that lives at the landfill? What about any groundwater under the facility?

5. Are you satisfied with the current treatment of local garbage? How could this treatment be improved?

6. What can citizens do to reduce solid wastes? What can YOU do?

7. What misconceptions did you have about landfills before the field trip? What surprised you about the facility?

I. Objectives

Environmental problems are receiving more and more public attention. After completing this chapter, the student will:

1. Realize that most environmental problems are complex and poorly understood.
2. Understand that solutions to environmental problems will not be easily, or cheaply, reached.
3. Gain an awareness of what each person can do to minimize damage to the environment.
4. Gain experience in writing and presenting a research paper.

II. Introduction

You cannot watch television or pick up a newspaper without hearing or reading about environmental problems and issues on a local, national, and international level. We are learning that our mineral and energy resources are not infinite, but will someday be depleted. The natural recycling of water and air is not sufficient to ensure the availability or purity of these resources for billions more people. As we double our population every forty years, we are putting increasing pressure upon the environment to meet our needs and desires. The infinite growth that entrepreneurs strive for cannot continue amidst finite resources.

Human beings must strike a balance between prosperity and sustainability if we are to remain a prominent species on the earth. The parameters of a sustainable society include recycling, efficiency, renewable resources, and zero population growth. Historically, humans have not worked within these limits. The challenge of the coming decades will be to reverse these trends to slow down the consumption and increase protection of our natural resources.

Forecasters disagree as to when our resources will be gone, but there is no question that it will happen. Should we deal with the problem now, or should we postpone action in the hope that new technology will provide a solution before our resources are gone? Can we afford to leave our future existence up to chance? The economics of depletion reason that as a resource becomes scarce, it becomes more valuable. Therefore, the remaining amount must be conserved, or new, expensive technology must be developed to extract the remainder more efficiently. However, human society is slow to change. A more likely scenario may be that we will go on consuming resources at ever-increasing rates until we reach a crisis. Through the hardship of our offspring, we may or may not find an alternative lifestyle. It is up to our generation to become motivated and ensure that this scenario does not happen.

III. Relevance: Developing Research Skills

As a person interested in environmental issues, it is important to have the skills necessary to research a topic in order to present the information in an understandable, precise manner. It is not enough to simply say "the environment needs protecting." When discussing an issue, you must have facts on hand, but, just as important, you must be able to convey ideas and issues in a clear manner. The first step is often researching your topic of interest. One key to a good research paper is to anticipate spending a significant amount of time gathering information, synthesizing it, and writing your paper. Group work requires coordination and cooperation, often working with people who have different schedules and time frames. Take these facts into account, and allocate enough time to do a good job.

IV. Activity

Working in groups of 2 to 3, students will research an environmental topic of their choice and present a 5- to 10-minute oral report on their research. Oral reports may be presented in debate style (with opposing views presented by the group), in reporting style, or any other style best suited to that topic. Make your oral report interesting and stimulating. Practice giving it before your real presentation. Be sure everyone in the group participates, and try to involve the audience. In short, be creative and have fun with your presentation. In addition, each group will turn in a typewritten paper, 3 to 5 pages long, covering its research.

Time will be given during this lab period to go to the library to start your research. However, additional time will be required to properly research and write this report, which should be typewritten and double-spaced. At least five sources should be used, with references properly documented at the end of the paper.

V. Procedure

Use the following lists as a starting point to determine what you will research and the places you can start looking for information. List A provides some research ideas, or you may choose one of your own liking. Lists B, C, and D include some of the best sources of recent information regarding many different environmental topics. Also, talk with your librarian about specific material your school may have.

LIST A: Topics

Deforestation
   ranching
   peasant migration
   First World impact
   multi-national corporations
   old-growth forests of North America

Energy
   alternative types
   renewable vs. nonrenewable forms
   projections for future uses
   conservation
   policy

Agriculture
   pesticides
   sustainable agriculture
   Third World dependency on First World economics
   the family farm
   loss of farmland
   erosion, desertification, salinization, irrigation

Conservation
   agriculture
   biological diversity
   habitat
         coral reef
         desert
         estuarine
         forest, etc.
   resources
         soil
         water
         air

Humans and Wildlife
   urban expansion vs. species extinction
   species extinction
   habitat loss
   ecotourism
   recreationists' impacts
   human dimensions of wildlife management
   animal rights
   wildlife management

Solid and toxic waste
   recycling
   Superfund sites

Environmental policy and laws
   political systems and resource management
   regulatory agencies and associations
   laws and Congressional Acts

Career potential in environmental positions

Environmental ethics
   celebrities and rock musicians
   ecofeminism

Environment and religion

Environment and the military

Environment and technology

Topics in the news
  global warming
  ozone depletion
  oil spills
  nuclear disasters, etc.

LIST B: Indices to Periodicals

Applied Science and Technology Index

Biological and Agricultural Index

Reader's Guide to Periodical Literature

Social Sciences Index

Government Repositories (located in many university libraries)

LIST C: Professional Periodicals

Advances in Environmental Science and Technology

Agricultural Water Management

Agriculture, Ecosystems, and Environment

American Journal of Alternative Agriculture

Aquaculture

The Auk

Biological Conservation

Chemical & Engineering News

The Condor

Conservation Biology

Environment and Behavior

Environmental Ethics

Environment and Planning

Environmental Pollution

Environmental Research

Environmental Science & Technology

Foreign Policy

Journal of Energy Resources Technology

Journal of Environmental Economics and Management

Journal of Environmental Health

Journal of Environmental Quality

Journal of Environmental Sciences

Journal of Hazardous Materials

Journal of the Institute of Energy

Journal of Social Issues

Journal of Soil and Water Conservation

Journal of Wildlife Management

Monthly Review

New Scientist

New Statesman and Society

Organic Gardening

Political Studies

Power

Soil and Water Conservation News

Water Research
Wildlife Society Bulletin
Wilson Bulletin

LIST D: General Periodicals

Animal's Agenda

Audubon Society

BioScience

Current Health

Discover Magazine

Environment

Field & Stream

Futurist

International Wildlife Society

Mother Earth News

Mother Jones

National Geographic

National Wildlife Society

Outdoor Life

Psychology Today

Science

Science News

Sierra Club

Smithsonian

The Nation

Time

U.S. News & World Report

USA Today

Utne Reader

World Health

World Press Review

VI. Questions

1. What factors influenced your choice of topic?

2. Was information easy or hard to find regarding your research issue? Why?

3. What information was the easiest to access (books, magazines, journals, etc.)? Which format was the easiest to understand? Do these two questions agree with each other? Why or why not?

4. Did your previous college experience help you conduct this research? What is the most important thing to know when conducting this type of research?

Field Trip Suggestions

1. Visit a local zoning board meeting or city council meeting when questions of land-use priorities are on the agenda. List the interest groups present at the meeting and describe the major points of view expressed by each group.
2. Visit a park and have a planner show how decisions were made regarding specific land uses. Select one portion of the park and list the reasons for developing it for its particular use.
3. Visit a supermarket and determine the origin of five fruits or vegetables, two types of fresh meats, one canned meat, one canned fish, one package of coffee, one package of tea, and one package of bread. Some of these items will have the origin printed on the label. To find the place of origin of some of the items, you may need to ask the produce manager or the butcher.

   Plot the places of origin on a map of the world. Select one domestic and one foreign item and list the steps necessary to get these items to market.
4. Visit a clothing store and determine the country of origin of five items by reading the labels. Use data from the *United Nations Data Books, The Population Reference Bureau Annual Data Sheet,* or other sources to determine typical wage rates for people in those countries.

   Plot the countries of origin on a map of the world. Write a paragraph describing why you think these items were manufactured where they were.
5. Visit your state, county, or city officials and discuss environmental legislation with them. Ask them to state their position on a locally important environmental issue. Ask them to list what environmentally significant legislation or ordinances they have supported in the past year.
6. Visit a local recycling center and list the kinds of materials recycled. Describe how the materials are processed within the center. Determine the market for the materials. Who uses the recycled materials, and how much are they willing to pay?
7. Visit a junkyard. Make a list of the kinds of materials accepted. Determine the price paid for each kind of material.
8. Attend a public meeting on an environmental issue.

Alternative Learning Activities

1. Invite your state or federal representatives or senators to address the class or school on environmental issues.
2. Write a letter outlining your position on a piece of environmental legislation to an appropriate government official.
3. Volunteer your time to participate in an environmentally significant activity.
4. Select a piece of land and do an environmental history of it. You may need to interview the current owners and search the records in the county register of deeds office.
5. Select a large piece of land and develop a map that shows suitable uses for various portions of it.
6. Determine what current environmentally significant legislation is before Congress (select one bill). How would passage of the bill affect your area? Keep a log that lists the progress of the bill.
7. Use a map of your local community. Locate all the open space that is available to the public (playgrounds, parks, golf courses, natural areas, and so forth). Do all regions of your community have equal access to open space? How does this relate to economic conditions, politics, and community planning?
8. Identify three places in your community that illustrate good planning and three that illustrate poor planning. Prepare visual aids that illustrate your points.
9. Determine how much water you use per day.
10. Pick up litter along a road or section of beach. Inventory the waste found. Identify causes of the litter and discuss solutions to the problem.

Random Numbers Table

24	39	65	81	94	82	77	6	0	4
87	13	19	49	71	37	26	78	91	16
55	0	11	41	4	14	22	70	52	38
14	47	96	54	85	11	70	13	100	65
43	65	15	74	72	44	74	66	14	57
77	16	42	36	49	59	3	85	0	40
47	77	100	43	57	45	60	88	34	23
60	58	41	57	89	38	70	76	54	20
44	27	27	77	8	47	68	37	20	10
13	71	3	100	54	25	68	48	6	14
59	95	79	99	47	5	56	70	11	9
4	59	33	83	99	100	55	51	7	62
11	26	98	77	0	58	37	80	98	18
20	89	74	85	12	47	20	60	48	33
55	79	50	93	7	42	33	83	43	40
38	66	40	93	82	67	75	72	43	38
29	32	86	11	71	20	50	7	31	92
82	84	18	93	54	76	99	42	53	23
93	77	70	27	83	93	72	50	27	37
100	79	50	90	70	74	100	99	18	33
58	92	58	80	77	31	11	12	54	44
71	12	84	98	45	74	23	29	62	29
11	18	71	27	59	13	25	67	69	94
74	25	90	76	61	74	71	45	52	48
25	31	78	38	55	9	37	78	53	27
28	5	44	52	54	56	64	46	86	79
48	48	12	28	25	81	62	14	94	57
41	80	76	6	54	89	4	89	59	60
76	33	33	77	2	48	80	84	84	52
74	95	64	74	60	37	70	63	14	46
2	65	77	52	77	97	61	46	90	90
55	87	4	18	23	38	20	89	16	78
43	19	93	95	5	60	80	92	43	4
56	15	10	33	94	53	10	50	46	90
29	33	4	73	58	71	29	80	92	27
62	51	80	57	33	59	17	65	68	80
68	4	10	60	91	3	59	20	68	36
40	98	97	13	21	78	70	6	64	92
72	87	52	63	39	95	66	61	7	19
56	35	82	89	84	17	80	84	74	35
49	46	59	43	31	97	53	10	79	58
30	60	74	65	57	21	76	88	52	73
40	24	9	39	29	3	68	21	68	59

To transform e^x to x, find the e^x on the appropriate graph, below, and locate the corresponding x value; if e^x is between 0 and 15, use graph B.1; if it is between 15 and 60, use graph B.2.

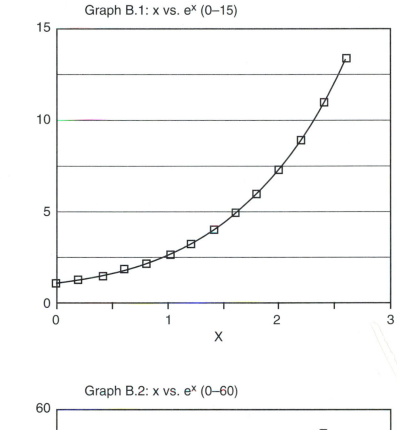

Graph B.1: x vs. e^x (0–15)

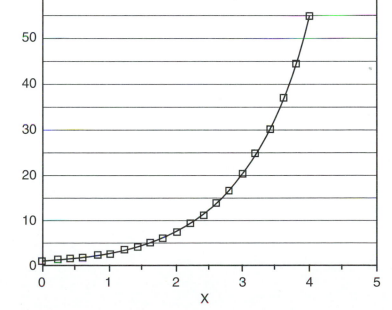

Graph B.2: x vs. e^x (0–60)

Code	Wind speed (mph)	Wind speed (kph)	Wind speed (knots)	Wave height (feet)	Designation	Description
0	<1	<1	<1	0	Calm	Smoke rises vertically; sea mirror-calm; tree leaves do not move
1	1–3	1–5	1–3	0	Light air	Smoke drift indicates wind direction; weather vane does not move
2	4–7	6–11	4–6	0–1/3	Light breeze	Wind felt on face; weather vane begins to move; leaves rustle; small wavelets
3	8–12	12–19	7–10	1/3–1+	Gentle breeze	Wind extends light flags; leaves and twigs in constant motion; large wavelets
4	13–18	20–28	11–16	2–4	Moderate breeze	Dust and loose paper raised; small branches move; small waves; many whitecaps
5	19–24	29–38	17–21	4–8	Fresh breeze	Small trees sway; moderate waves
6	25–31	39–49	22–27	8–13	Strong breeze	Large branches move; wind whistles in wires; larger waves forming
7	32–38	50–61	28–33	13–20	Moderate gale	Whole trees move; walking affected
8	39–46	62–74	34–40	13–20	Fresh gale	Twigs break off trees; walking difficult; moderately high waves
9	47–54	75–88	41–47	13–20	Strong gale	Slight structural damage occurs; high waves; branches break
10	55–63	89–102	48–55	20–30	Whole gale	Trees uprooted; considerable structural damage; very high waves w/overhanging crest
11	64–75	103–117	56–63	30–45	Storm	Widespread damage; extremely high waves; sea covered w/white foam patches
12	75+	117+	64+	45+	Hurricane	Severe and extensive damage; visibility of sea greatly reduced